A PLAN FOR A
RESEARCH PROGRAM ON

Aerosol Radiative Forcing and Climate Change

Panel on Aerosol Radiative Forcing and Climate Change
Board on Atmospheric Sciences and Climate
Commission on Geosciences, Environment, and Resources
National Research Council

NATIONAL ACADEMY PRESS
Washington, D.C.   1996

AN EXAMPLE OF A REGIONAL AEROSOL DISTRIBUTION

This NASA shuttle photograph (opposite page), taken on a March morning in 1994, and the explanatory diagram show a regional "haze" inland from the coast of California. While no chemical analysis is available, the haze is a widespread aerosol from sources that probably include smoke particles from biomass combustion and cities in the region. The enhanced albedo due to the haze causes sunlight to be reflected upward and thereby to fail to reach the ground. This constitutes a "direct climate forcing."

The aerosol cloud is visible from the northern extremity of the Sacramento Valley on the left to the Bakersfield area of the San Joaquin Valley on the right, a distance of about 600 kilometers. The Sierra Nevada mountains and the coastal range bound the aerosol-laden valley.

The photograph also shows coastal stratus clouds that extend along the coast and penetrate into the San Francisco Bay region. The albedo of these clouds, which can be influenced by anthropogenic aerosols, clearly controls the albedo of the oceanic portion of this view.

(Shuttle photograph SS062-86-066, courtesy of the Earth Science Branch, NASA/ Johnson Space Center, Houston, Texas)

Cover art by Carrie Mallory. Ms. Mallory received her Bachelor of Fine Arts degree from the Cooper Union. She draws on the natural world and the effects of age on manmade objects for many of her themes. She has exhibited at a number of juried shows in the Northern Virginia area and has provided art for several NRC report covers. The art for this cover involved transferring an original photograph to an already cracked lithograph stone and adding texture with traditional litho crayons. As expected, the stone cracked further during the printing process, yielding only a few prints before disintegrating completely.

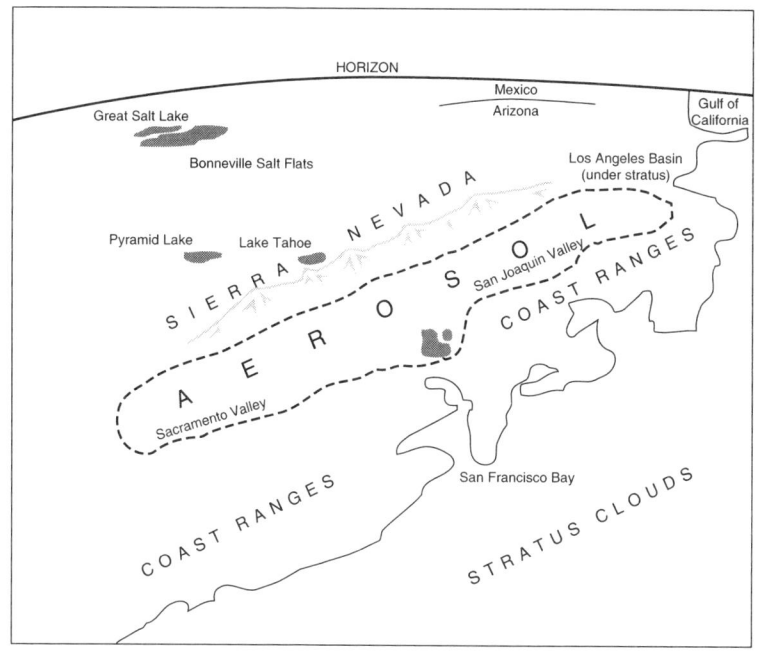

National Academy Press • 2101 Constitution Avenue, N.W. • Washington, D.C. 20418

NOTICE: The project that is the subject of this report was approved by the Governing Board of the National Research Council, whose members are drawn from the councils of the National Academy of Sciences, the National Academy of Engineering, and the Institute of Medicine. The members of the committee responsible for the report were chosen for their special competences and with regard for appropriate balance.

This report has been reviewed by a group other than the authors according to procedures approved by a Report Review Committee consisting of members of the National Academy of Sciences, the National Academy of Engineering, and the Institute of Medicine.

The National Academy of Sciences is a private, nonprofit, self-perpetuating society of distinguished scholars engaged in scientific and engineering research, dedicated to the furtherance of science and technology and to their use for the general welfare. Upon the authority of the charter granted to it by the Congress in 1863, the Academy has a mandate that requires it to advise the federal government on scientific and technical matters. Dr. Bruce M. Alberts is president of the National Academy of Sciences.

The National Academy of Engineering was established in 1964, under the charter of the National Academy of Sciences, as a parallel organization of outstanding engineers. It is autonomous in its administration and in the selection of its members, sharing with the National Academy of Sciences the responsibility for advising the federal government. The National Academy of Engineering also sponsors engineering programs aimed at meeting national needs, encourages education and research, and recognizes the superior achievements of engineers. Dr. Harold Liebowitz is president of the National Academy of Engineering.

The Institute of Medicine was established in 1970 by the National Academy of Sciences to secure the services of eminent members of appropriate professions in the examination of policy matters pertaining to the health of the public. The Institute acts under the responsibility given to the National Academy of Sciences by its congressional charter to be an adviser to the federal government and, upon its own initiative, to identify issues of medical care, research, and education. Dr. Kenneth I. Shine is president of the Institute of Medicine.

The National Research Council was organized by the National Academy of Sciences in 1916 to associate the broad community of science and technology with the Academy's purposes of furthering knowledge and advising the federal government. Functioning in accordance with general policies determined by the Academy, the Council has become the principal operating agency of both the National Academy of Sciences and the National Academy of Engineering in providing services to the government, the public, and the scientific and engineering communities. The Council is administered jointly by both Academies and the Institute of Medicine. Dr. Bruce M. Alberts and Dr. Harold Liebowitz are chairman and vice chairman, respectively, of the National Research Council.

Support for this project was provided by the Department of Agriculture, the Department of Energy, the Environmental Protection Agency, the Office of Naval Research of the Department of Defense, the Air Force Office of Scientific Research, the National Aeronautics and Space Administration, the National Oceanic and Atmospheric Administration, and the National Science Foundation under Grant No. ATM-9316824. Any opinions, findings, and conclusions or recommendations expressed in this publication are those of the author(s) and do not necessarily reflect the views of the above-mentioned agencies.

Library of Congress Catalog Card Number 96-67382
International Standard Book Number 0-309-05429-X

Additional copies of this report are available from:

National Academy Press
2101 Constitution Avenue, NW
Box 285
Washington, DC 20055
800-624-6242
202-334-3313 (in the Washington Metropolitan Area)
B-705

Copyright 1996 by the National Academy of Sciences. All rights reserved.

Printed in the United States of America

PANEL ON AEROSOL RADIATIVE FORCING AND CLIMATE CHANGE

JOHN H. SEINFELD (*Chair*), California Institute of Technology, Pasadena
ROBERT CHARLSON, University of Washington, Seattle
PHILIP A. DURKEE, Naval Postgraduate School, Monterey, California
DEAN HEGG, University of Washington, Seattle
BARRY J. HUEBERT, University of Hawaii, Honolulu
JEFFREY KIEHL, National Center for Atmospheric Research, Boulder, Colorado
M. PATRICK MCCORMICK, Langley Research Center, National Aeronautics and Space Administration, Hampton, Virginia
JOHN A. OGREN, Climate Monitoring and Diagnostics Laboratory, National Oceanic and Atmospheric Administration, Boulder, Colorado
JOYCE E. PENNER, Lawrence Livermore National Laboratory, Livermore, California
VENKATACHALAM RAMASWAMY, Geophysical Fluid Dynamics Laboratory, National Oceanic and Atmospheric Administration, Princeton, New Jersey
W. GEORGE N. SLINN, Pacific Northwest Laboratories, Richland, Washington

Staff

DAVID H. SLADE, Senior Program Officer
DORIS BOUADJEMI, Administrative Assistant

BOARD ON ATMOSPHERIC SCIENCES AND CLIMATE

JOHN A. DUTTON (*Chair*), Pennsylvania State University, University Park
ERIC J. BARRON, Pennsylvania State University, University Park
WILLIAM L. CHAMEIDES, Georgia Institute of Technology, Atlanta
CRAIG E. DORMAN, Department of Defense, Washington, D.C.
FRANCO EINAUDI, Goddard Space Flight Center, Greenbelt, Maryland
MARVIN A. GELLER, State University of New York, Stony Brook
PETER V. HOBBS, University of Washington, Seattle
WITOLD F. KRAJEWSKI, The University of Iowa, Iowa City
MARGARET A. LEMONE, National Center for Atmospheric Research, Boulder, Colorado
DOUGLAS K. LILLY, University of Oklahoma, Norman
RICHARD S. LINDZEN, Massachusetts Institute of Technology, Cambridge
GERALD R. NORTH, Texas A&M University, College Station
EUGENE M. RASMUSSON, University of Maryland, College Park
ROBERT J. SERAFIN, National Center for Atmospheric Research, Boulder, Colorado

Staff

WILLIAM A. SPRIGG, Director
H. FRANK EDEN, Senior Program Officer
MARK D. HANDEL, Senior Program Officer
DAVID H. SLADE, Senior Program Officer
ELLEN F. RICE, Reports Officer
DORIS BOUADJEMI, Administrative Assistant
THERESA M. FISHER, Administrative Assistant
MARK BOEDO, Project Assistant

COMMISSION ON GEOSCIENCES, ENVIRONMENT, AND RESOURCES

M. GORDON WOLMAN (*Chair*), The Johns Hopkins University, Baltimore, Maryland
PATRICK R. ATKINS, Aluminum Company of America, Pittsburgh, Pennsylvania
JAMES P. BRUCE, Canadian Climate Program Board, Ottawa, Ontario
WILLIAM L. FISHER, University of Texas, Austin
JERRY F. FRANKLIN, Unviersity of Washington, Seattle
GEORGE M. HORNBERGER, University of Virginia, Charlottesville
DEBRA KNOPMAN, Progressive Foundation, Washington, D.C.
PERRY L. MCCARTY, Stanford University, California
JUDITH E. MCDOWELL, Woods Hole Oceanographic Institution, Massachusetts
S. GEORGE PHILANDER, Princeton University, New Jersey
RAYMOND A. PRICE, Queen's University at Kingston, Ontario
THOMAS C. SCHELLING, University of Maryland, College Park
ELLEN SILBERGELD, University of Maryland Medical School, Baltimore
STEVEN M. STANLEY, The Johns Hopkins University, Baltimore, Maryland
VICTORIA J. TSCHINKEL, Landers and Parsons, Tallahassee, Florida

Staff

STEPHEN RATTIEN, Executive Director
STEPHEN D. PARKER, Associate Executive Director
MORGAN GOPNIK, Assistant Executive Director
GREGORY SYMMES, Reports Officer
JAMES MALLORY, Administrative Officer
SANDI FITZPATRICK, Administrative Associate
SUSAN SHERWIN, Project Assistant

Foreword

As this report was receiving its final editing, Working Group I of the Intergovernmental Panel on Climate Change released its *Summary for Policy Makers* (IPCC, 1995b). The first section of the IPCC summary ("Greenhouse gas concentrations have continued to increase") documents the increase of greenhouse gases with arguments that are now almost universally accepted in the scientific community. The second section ("Anthropogenic aerosols tend to produce negative radiative forcings") quantifies the "direct" negative forcing of anthropogenic aerosols as a global average of 0.5 watts per square meter, and suggests that there is also an "indirect" negative forcing of a similar magnitude. The remainder of the IPCC summary presents evidence that supports its view that aerosol radiative forcing plays a fundamental role in global climate change. The National Research Council's Panel on Aerosol Radiative Forcing and Climate Change agrees with the IPCC findings.

The United States has taken a leading role in investigating the aerosol effect. Recent federal funding, at a level of about one-half percent of the U.S. Global Change Research Program, has supported efforts to provide preliminary estimates of the mechanisms, magnitudes, uncertainties, and environmental consequences of aerosol radiative forcing. As the following report points out, however, there is much to be done before the scientific community can confidently advise those charged with developing policy and legislation on the significance and timing of this climate-perturbing problem. For example, currently even the composition and the spatial patterns of aerosol distribution are, in large part, tentative due to the paucity of basic measurements. Model descriptions of the process of aerosol formation, the environmental behavior of aerosols, and their effect on the dynamics of climate are all somewhat conjectural. The research that has been carried out in this country and abroad, however, is sufficient to support the main findings of the IPCC Working Group I and this panel: that aerosol radiative forcing of climate is not only an interesting scientific issue but also is likely to play a significant role in our future climate.

John Seinfeld
Chair
Panel on Aerosol Radiative
Forcing and Climate Change

Contents

EXECUTIVE SUMMARY   1

1 CLIMATE FORCING BY AEROSOLS   7
   Atmospheric Aerosols   8
   Aerosol Radiative Forcing of Climate   11
   Evidence for Radiative Forcing by Anthropogenic Aerosols   13
      Direct Forcing   13
      Indirect Forcing   17
   Evidence for Climate Response to Anthropogenic Aerosol
      Forcing   20
   Radiative Forcing of Climate by Stratospheric Aerosols   23
      Evidence for Climate Response to Stratospheric Aerosol
         Perturbation   25
      Inferences from Stratospheric Aerosol Research   26
   Climate Forcing by Key Tropospheric Aerosol Types   26
   Conclusions   29

2 ELEMENTS OF A RESEARCH PROGRAM FOR AEROSOL
   FORCING OF CLIMATE   35
   Global Climate Models   37
      Atmospheric General Circulation Models   38
      Atmospheric Chemical Transport Models   38
      Recommended Research on Global Climate Modeling of Aerosol
         Radiative Forcing   42

Process Research 43
    Optical Properties 43
    Recommended Process Research on Aerosol and Cloud
        Optical Properties 47
    Aerosol Dynamics 47
    Recommended Process Research on Aerosol Dynamics 51
    Aerosol Sinks 52
    Recommended Process Research on Aerosol Sinks 58
    Aerosols and Ice Formation in Clouds 59
    Recommended Process Research on Aerosols and Ice Formation
        in Clouds 61
    Aerosol Process Models 62
    Recommended Process Research on Aerosol Models 64
Field Studies 64
    Closure Experiments 65
    Multiplatform Field Campaigns 68
    Recommended Field Studies 71
Satellite Observations and Continuous In Situ Monitoring 71
    Satellite Remote Sensing of Aerosols 73
    In Situ Monitoring of Aerosols 78
    Recommended Surface-Based Monitoring Programs 83
    Mobile Platforms 83
    Recommended Mobile Monitoring Programs 85
Recommended Technology Developments 85
Summary of a Research Program on Aerosol Forcing of Climate 88

3   SENSITIVITY/UNCERTAINTY ANALYSIS AND
     THE SETTING OF PRIORITIES   89
     Integration via Sensitivity/Uncertainty Analyses 90
     An Example of Sensitivity Analysis: Direct Radiative Forcing 90
     Frameworks for Research and Funding Priorities 94
     An Example of Sensitivity and Uncertainty Analyses 98
     Summary 104

4   THE PROPOSED ICARUS PROGRAM AND RECOMMENDED
     RESEARCH   107
     The ICARUS Strategy 108
     Organizational Structure of the ICARUS Program 111
     Research Program 112
        Global Climate Model Development 114
        Process Research 115
        Multiplatform Field Campaigns 117

Satellite System Development　119
　　　System Integration and Assessment　120

REFERENCES　123

APPENDIX A　ILLUSTRATIONS OF RECOMMENDED RESEARCH,
　　　　　　　EMPHASIZING RECENT LITERATURE　135

APPENDIX B　ACRONYMS AND OTHER INITIALS　159

List of Tables

Table 1.1 Comparison of Climate Forcing by Aerosols with Forcing by Greenhouse Gases (GHGs): Fundamental Differences in Approach to Determination and Nature of Forcing 14
Table 1.2 Source Strength, Atmospheric Burden, Extinction Efficiency, and Optical Depth for Various Types of Aerosols 28
Table 1.3 Estimates of Direct Climate Forcing (W m^{-2}) by Anthropogenic Aerosols 30
Table 1.4 Key Anthropogenic Aerosol Types, Associated Forcing Mechanisms, and Status of Understanding 32
Table 2.1 Global Models Currently Used to Study Aerosol Forcing: (A) Atmospheric General Circulation Models for Aerosol Forcing Calculations; (B) Global and Synoptic Models for Chemical Transport of Aerosols 39
Table 2.2 Satellite Instruments 76
Table 2.3 Aerosol Properties Needed at Continuous Monitoring Sites 80
Table 2.4 Categories of Sites to Monitor Intensive Properties 81
Table 3.1 Sensitivity Calculations for a Sulfate Aerosol Layer Below Clouds 92
Table 3.2 Sensitivity Calculations for an Aerosol Layer Above Lowest Cloud Layer 93
Table 3.3 Factors Contributing to Estimates of the Direct Forcing by Anthropogenic Sulfate (A) and Biomass Burning (B) Airborne Particles, Estimated Ranges, and Resulting Uncertainty Factors (for estimates of changes in reflected solar radiation) 100

List of Figures

Figure 1	Organizational structure of the ICARUS program.	3
Figure 1.1	Estimated Northern Hemisphere and regional anthropogenic sulfur emissions over the past century.	10
Figure 2.1	General components of an integrated aerosol-climate research program.	36
Figure 2.2	Direct and indirect forcing mechanisms associated with sulfate aerosols.	44
Figure 2.3	Observations of continental haze by LITE (Lidar In-Space Technology Experiment).	75
Figure 2.4	Ship tracks off the coast of Northern California.	84
Figure 3.1	Sensitivity of aerosol forcing for an aerosol layer below cloud.	92
Figure 3.2	Sensitivity of aerosol forcing for an aerosol layer above lowest cloud layer.	93
Figure 3.3	Qualitative indications of current radiative forcing uncertainties for indirect effects (separately for marine and continental clouds) and for direct effects (separately for organic and inorganic aerosols) and a qualitative indication of the uncertainty goal (to be defined by USGCRP) for the first phase of ICARUS research.	95
Figure 3.4	Qualitative indication of relative ICARUS research priorities for different topics, with the differences from Figure 3.3 resulting from weighting the uncertainties of Figure 3.3 with USGCRP "strategic" and "integrating" priorities; here, the weighting has been by assumed amounts.	96

Figure 3.5	Qualitative indication of relative funding priorities (resource allocations) for the indicated broad research topics, with the differences from Figure 3.4 (research priorities) resulting from weighting these research priorities with costs to perform the research; here, the weighting has been by assumed amounts.	97
Figure 3.6	Plot of the uncertainties listed in Table 3.3A for sulfate aerosols, with a qualitative indication of the level to which the uncertainty could be set as a goal for the first phase of ICARUS research.	102
Figure 3.7	Qualitative indication of research priorities for direct radiative effects of sulfate aerosols, derived from Figure 3.6 (uncertainties) by weighting with such factors as mentioned in the text.	103
Figure 3.8	Qualitative indication of the relative costs to reduce the uncertainties shown in Figure 3.6, consistent with the research priorities shown in Figure 3.7, accounting for the cost of performing the research (e.g., a prorated portion of satellite costs to measure backscattered radiation).	103
Figure 3.9	Qualitative indication of the relative contributions from different processes to current uncertainty in the atmospheric lifetime of aerosol sulfate, with a qualitative indication of the level to which the uncertainty could be set as a goal for the first phase of ICARUS research.	104
Figure 3.10	Qualitative indication of funding priorities to reduce the uncertainties shown in Figure 3.9.	105
Figure 4.1	Organizational structure of the ICARUS program.	112

Executive Summary

It has been more than a century since scientists first predicted that changes in the chemical composition of the atmosphere, particularly increasing concentrations of carbon dioxide from anthropogenic activities, might change the Earth's heat balance and cause a warming of the atmosphere (Arrhenius, 1896). In contrast, it is only recently that scientists began to consider quantitatively how anthropogenic aerosols—very small particles that are suspended in the air—affect global climate (Charlson et al., 1990). Although there is much evidence to suggest that aerosols cause cooling, it is not confidently known just how large such an effect might be. Because the effects of atmospheric aerosols on climate are still poorly understood, and because these effects could play an essential role in explaining past climate trends and in predicting future climate change, the results of aerosol research will likely be highly relevant to current international policy activities involving the United States.

To approach this complex subject systematically, the scientific community has divided the problem into two major parts referred to as climate *forcings* and climate *responses*. Climate forcings are the changes in the energy balance of the Earth's environmental system that are imposed upon it. Forcings are calculated or measured in units of heat flux—watts per square meter ($W\ m^{-2}$). Climate responses are the meteorological results of these forcings, including changes in temperature, wind, rainfall, and their probability distributions.

Aerosol forcings are divided into two types: *direct* and *indirect*. Direct forcing relates to the direct interaction of aerosol particles with the incoming solar radiation in, essentially, non-cloudy conditions. Indirect forcing relates to forcing under cloudy conditions, during which the aerosol and the cloud may interact in a number of ways: the aerosol may increase the cloud droplet concentration, thereby influencing the cloud albedo; the aerosol may influence cloud persistence; or it may reduce the possibility of precipitation. Even modestly confident quantification of indirect forcing is not yet possible.

CHARGE TO THE PANEL

In their letter to the chair of the Board on Atmospheric Sciences and Climate (BASC) of the National Research Council, the director of the Environmental Sciences Division, Department of Energy; the chief of the Climate and Hydrologic Systems Branch, National Aeronautics and Space Administration; the director of the Office of Global Programs, National Oceanic and Atmospheric Administration; and the director of the Division of Atmospheric Sciences, National Science Foundation, requested that

> In view of the potential significance of climate forcing by both man-made and natural aerosols and the relatively undeveloped ability to describe this forcing in climate models, we are writing to request that BASC advise the government on development of a strategy and potential program plan for a U.S. research effort. Given current agency priorities concerning climate and global change, we feel that the appropriate scientific focus for the U.S. program of aerosol research is the climatic effects of aerosol particles . . .

(see Chapter 1 for the complete Statement of Task). This report summarizes current understanding of the effects of aerosols on climate change and recommends a research program to identify and prioritize the research required to determine the effects of aerosol forcing on the atmosphere's energy balance.

FINDINGS

The panel's main findings are that (1) anthropogenic aerosols reduce the amount of solar radiation reaching the Earth's surface, (2) anthropogenic aerosols provide a negative climate forcing function for large regions, (3) global models suggest that sulfate aerosols produce a direct forcing in the Northern Hemisphere of the same order of magnitude as that from anthropogenic greenhouse gases, but opposite in sign, and (4) there is substantial uncertainty about the magnitude and spatial distribution of the radiative forcing by aerosols. Reduction in this uncertainty requires a scientifically and administratively integrated research program that could organize the research capabilities of the atmospheric science community in this country and permit cooperation with other national and international research programs. The following sections summarize the management strategy and specific research projects recommended to achieve this goal.

MANAGEMENT RECOMMENDATIONS

Action 1: Establish leadership by empowering an Interagency Climate-Aerosol Radiative Uncertainties and Sensitivities (ICARUS) Program (see Figure 1 for the proposed ICARUS organizational structure).

Action 2: Mobilize ICARUS by developing methods to define research priorities.

Action 3: Develop a multiagency, integrated research program.

Action 4: Maintain ICARUS's leadership in aerosol research by applying steadily improving sensitivity analysis to research designs.

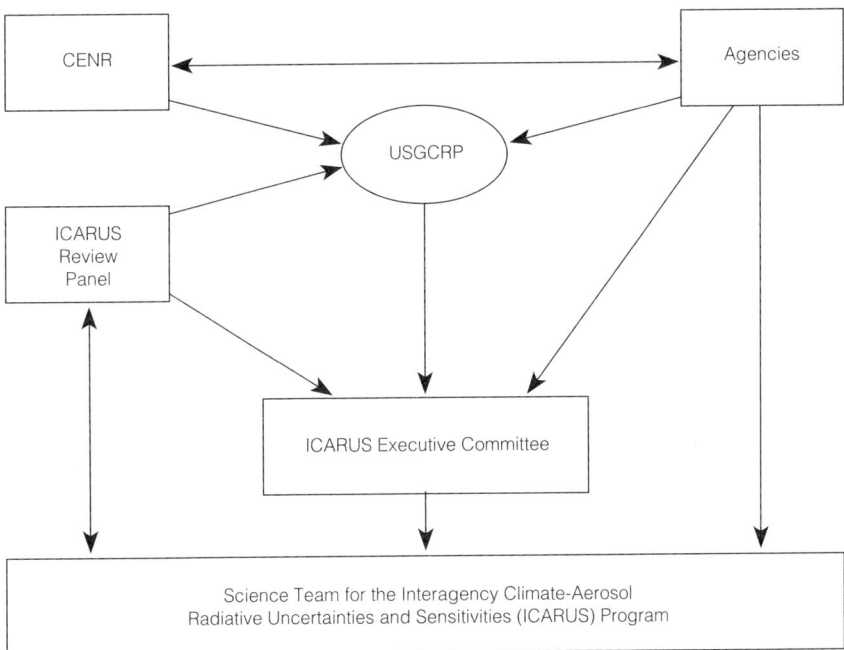

FIGURE 1 Organizational structure of the ICARUS program.
NOTE: CENR = Committee on Environment and Natural Resources.

TECHNICAL RECOMMENDATIONS

It is recommended that a research program be developed to reduce uncertainties in calculated aerosol forcing at the top of the atmosphere to within ±15 percent both globally and locally. This limit of uncertainty is equivalent to that required in estimating the greenhouse gas forcing (IPCC, 1995a). Locally, this would imply an uncertainty in forcing of less than ±1.5 W m^{-2} (by assuming a local aerosol effect of -10 W m^{-2} in the diurnal average).

Six categories of research are recommended to achieve this goal:

1. global climate model development;
2. process studies (modeling, small-scale field studies) and technology development (excluding satellite development);
3. field studies: large, international field studies and systematic ship- and aircraft-based surveys of aerosol properties;
4. satellite observations and continuous measurements from surface-based research measurement networks;
5. technology development; and
6. system integration of the program (continuous coupling of modeling and measurement programs with ongoing assessment of sensitivities and uncertainties).

The following specific research undertakings are recommended for each of these categories.

SCIENCE DEVELOPMENT

Global Climate Model Development

- *Sensitivity and Uncertainty of Aerosol Forcing in Global Climate Models*: Using existing global climate and chemistry models, determine the global and local sensitivity of climate predictions to uncertainty in aerosol forcing and associated factors.
- *Development of Global-Scale Aerosol Radiative Forcing Models*: Develop aerosol radiative forcing models that incorporate chemical, physical, and meteorological processes for regional to global scales.

Process Research

- *Aerosol Formation and Growth by Nucleation and Gas-to-Particle Conversion*: Using field, laboratory, and theoretical approaches, determine the extent to which aerosol formation and growth by homogeneous nucleation and gas-to-particle conversion is occurring in the atmosphere.

- *Aerosol and Cloud Optical Properties*: Using field, laboratory, and theoretical approaches, determine the extent to which theoretical treatments of aerosol and cloud optical properties are applicable to ambient aerosols and clouds.
- *The Aerosol-Cloud Condensation Nuclei (CCN)-Cloud Drop Number Concentration (CDNC)-Cloud Albedo Linkage*: This project will develop an understanding of "indirect" aerosol forcing that depends on information on the relationship among aerosol number concentration, CCN, CDNC, and cloud albedo.
- *Aerosol Sinks*: Quantification of the rate of removal of aerosols from the atmosphere will be developed from field studies of particle size-specific dry deposition velocities over a variety of terrain, from precipitation scavenging measurements, and from storm-venting experiments.

Field Studies and Continuous In Situ Measurements

- *Multiplatform Field Campaigns*: The goal of these field studies is to integrate satellite radiation measurements and surface-based column-integrated radiation measurements and to quantitatively understand the processes controlling the formation, transformation, and removal of aerosols in different environments.
- *Mobile Platforms*: It is proposed that surface measurements be supplemented with systematic observations from a suitably instrumented aircraft, from ships, and from a balloon program, the primary uses of which would be to supply "ground truth" measurements for comparison with satellite observations and to provide typical values and variabilities of aerosol properties used in the models.
- *Surface-Based Stations for Continuous Research Measurement of Aerosols*: Establish a dual-density network of surface-based stations for the research measurement of characteristics of optical depth (dense network) and radiative, chemical, and microphysical properties of aerosols (low-density network).

Satellite Observations

- *Satellite Remote Sensing of Aerosols*: The primary objective of this project is to provide a geographically and vertically resolved climatology of aerosol extinction throughout the troposphere and stratosphere.

TECHNOLOGY DEVELOPMENT

Non-Satellite Technology Development

- *Instrumentation Development*: The purpose of this project is the development of instrumental systems intended specifically for aerosol forcing/climate research. Instruments include aerosol optical instrumentation, a CCN spectrometer, and in situ chemical analysis samplers.

Satellite Technology Development

- *Development of Extinction Profile Measurement Capability*: The goal is to develop the capability to measure extinction profiles (vertically resolved aerosol optical depths) on a global scale from a satellite.

SYSTEM INTEGRATION AND ASSESSMENT

The following projects are necessary to assure that the preceding technical recommendations are integrated and coordinated across the program, and that the results of the research are communicated promptly and authoritatively to those requiring technical guidance for decisionmaking.

- *Coordination*: The goal is to maintain communication among projects regarding the technical details of the ongoing efforts. Included are small projects for linking the research efforts, tracking projects, and reporting research results.
- *Integration and Assessment*: This small project will organize the data on aerosol forcing generated by the United States into formats suitable for presentation to the Intergovernmental Panel on Climate Change and other scientific and administrative bodies with differing requirements.
- *Integration of U.S. Research on Aerosol Forcing of Climate*: While some aspects of the U.S. research of aerosol forcing of climate will be supported directly by the funding agency members of ICARUS, there are others that may not be or that will require some additional coordination. Typically, these efforts include the maintenance of data quality assurance efforts, the construction of program-wide data bases, model and measurement intercomparison studies, and the support of meetings.

1

Climate Forcing by Aerosols

It is easy to take for granted the constancy of Earth's climate and the ability of humans to cope effectively with its natural variability. After all, the human race has survived ice ages and, more recently the Little Ice Age; the effects of large volcanic eruptions; and myriad events of violent weather, floods, and droughts. However, this oft-held assumption is misleading in its naivete and may even be dangerous. Recent advances in atmospheric science have shown that

- the chemical composition of the entire atmosphere of the planet has been changed due to human activity;
- these changes in gases and airborne particles result in changes in the heat balance of the planet; and
- the meteorological processes that, when taken together, constitute climate are dependent on and driven by local variations in this energy balance.

Thus, there are fundamental reasons to believe that changes in climate may result from these human-induced changes in atmospheric composition. The magnitude of these changes in climate, however, are poorly known.

In order to approach this complex subject systematically, the scientific community has divided the problem into two major parts, referred to as climate forcings and climate responses. Climate forcings are changes in the energy balance of the Earth that are imposed upon it; forcings are calculated or measured in units of heat flux—watts per square meter ($W\ m^{-2}$). Responses are the meteorological results of these forcings, including a large number of variable factors such as temperature, wind, rainfall, the probabilistic distribution of these, and extremes of weather.

The changes in energy balance as a result of changes in composition are calculated to be a few watts per square meter; some forcings are positive and others negative. Changes of this magnitude are a small, but finite, fraction of the average total energy flux into and out of the Earth's surface, which is approximately 200 W m^{-2}. Nonetheless, changes of a percent or two in heat flux are calculated in some climate models to produce significant changes in key meteorological parameters such as temperature and rainfall.

In order to be confident in our ability to predict future climate, we must accurately quantify these forcings, use them to predict the response of the climate system, and verify that the response is accurate through comparison to data (e.g., the historical temperature record). Whereas observing systems are now in place and adequate for quantifying greenhouse forcings, those for quantifying the forcing by anthropogenic aerosols do not exist. Further, the inclusion of aerosols and their effects in climate models is highly simplified and may contain errors. An incorrect or uncertain calculation of forcing by anthropogenic aerosols could significantly alter the understanding of climate response because of the requirement that the climate response to any given forcing should not be much different from that allowed by comparison to historical data.

Thus, a great deal is at stake in correctly and accurately quantifying all climate forcings. Without this, it will be impossible to develop any reliable predictive capacity for climate responses.

In this report, we consider just one family of forcings—those resulting from aerosols. We conclude that there is substantial evidence that these forcings are significant compared to those by greenhouse gases (GHGs), but that the current quantification of them is much more uncertain than for forcings by GHG. We further conclude that the ability to adequately predict climate change will depend on reducing these uncertainties through an integrated research program.

ATMOSPHERIC AEROSOLS

An aerosol is defined as a suspension of solid or liquid particles in a gas. Atmospheric aerosols are ubiquitous and often observable by eye as dust, smoke, and haze. Particles comprising the atmospheric aerosol range in sizes from nanometers (nm) to tens of micrometers (μm), that is, from large clusters of molecules to visible flecks of dust. Most of the smallest particles (less than about 0.1 μm) are produced by condensation, either from reactive gases in the atmosphere (e.g., sulfur dioxide) or in high-temperature processes (e.g., fire). Particles larger than about 1 μm are usually produced mechanically (windblown soil, sea spray, etc.). As a result of myriad production processes, atmospheric aerosol chemical composition is highly variable, with respect both to size and to spatial and temporal distribution. Because of different sizes and chemical compositions, aerosol particles have a wide range of lifetimes in air: from minutes to hours for the largest dust particles, from days to weeks for submicrometer smoke and haze particles in the troposphere, and up to two years for volcanic aerosol in the stratosphere.

There are two layers of the lower atmosphere: the troposphere and the stratosphere (i.e., below and above the tropopause at about 10-kilometer (km) altitude, respectively). Tropospheric aerosols vary significantly in amount and composition by region, with a characteristic horizontal spatial scale of variation ranging from 1 km to a few thousand kilometers. Notable examples of regionally defined tropospheric aerosol types are marine aerosol, industrial haze, desert dust, and smoke from biomass combustion. Because of its much longer residence time, stratospheric aerosol is substantially more homogeneous chemically and spatially than tropospheric aerosol. The presence of aerosols in the stratosphere is most evident following large volcanic eruptions. For example, the June 1991 eruption of Mt. Pinatubo in the Philippine Islands caused brilliant sunsets and sunrises worldwide through most of 1992.

A substantial fraction of today's tropospheric aerosol is *anthropogenic*. The highly visible haze that persists in all of the industrialized regions of the world consists mainly of sulfate and organic compounds from emissions of sulfur dioxide, organic gases, and smoke from biomass combustion. These emissions have changed dramatically over the past century; SO_2 emissions now amount globally to approximately 65-80 teragrams (Tg) (as elemental sulfur) per year, mainly from burning of solid fuels and smelting of metal ores. Figure 1.1 charts the growth of Northern Hemisphere sulfur emissions during the past century and, for comparison, provides estimates of Northern Hemisphere natural fluxes. [Southern Hemisphere anthropogenic emissions are perhaps 10 percent of those in the Northern Hemisphere.] Because the time scale for tropospheric air transport across the equator is much longer than that for the removal of sulfur compounds from the air and because most anthropogenic sulfur emissions are located in the Northern Hemisphere, most of the anthropogenic sulfate aerosol is found in the Northern Hemisphere. As a result, anthropogenic sulfur emissions are estimated to contribute about 80 to 90 percent of the total burden of sulfate aerosol in the Northern Hemisphere.

In their letter to the chair of the Board on Atmospheric Science and Climate (BASC) of the National Research Council, the director of the Office of Global Programs of the National Oceanic and Atmospheric Ad-

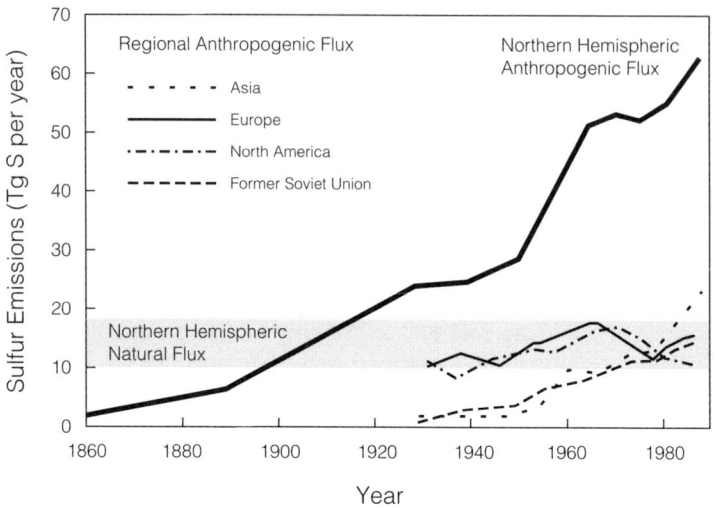

FIGURE 1.1 Estimated Northern Hemisphere and regional anthropogenic sulfur emissions over the past century. SOURCE: Dignon and Hameed (1989); Dignon and Gene (1995).

ministration (NOAA), the director of the Environmental Sciences Division of the U.S. Department of Energy (DOE), the director of the Division of Atmospheric Sciences of the National Science Foundation (NSF), and the chief of the Climate and Hydrologic System Branch of the National Aeronautics and Space Administration (NASA) wrote the following:

> Recent work has suggested that anthropogenic aerosols, especially sulfates, may exert a substantial radiative forcing of climate, comparable in magnitude, but opposite in sign, to the forcing from anthropogenic greenhouse gases. They may also play an important secondary role in climate as a source of cloud condensation nuclei. If these hypotheses are correct, they would have major implications on our perceptions of the entire climate change issue.
>
> In view of the potential significance of climate forcing by both manmade and natural aerosols and the relatively undeveloped ability to describe this forcing in climate models, we are writing to request that BASC advise the government on development of a strategy and potential program plan for a U.S. research effort. Given current agency priorities concerning climate and global change, we feel that the appropriate scientific focus for the U.S. program of aerosol research is on the climatic effects of aerosol particles . . . .
>
> Specifically, in our judgment, such a research program should be directed toward quantifying the radiative properties, sources and sinks of

tropospheric and stratospheric aerosols, and describing their forcing accurately in global models so that their climatic consequences may be evaluated. In our view, initial efforts should be directed at quantifying this forcing, with the objective of reaching an early understanding of its magnitude averaged over various geographical scales.

It also seems prudent that actions be identified so that these can be initiated right away. Such actions might include the planning of airborne field campaigns in the near-term as well as initiation of the design of satellite systems for characterizing the aerosol forcing, particularly for tropospheric aerosols, in view of the long lead times for developing such systems and for securing flight opportunities. Similarly, BASC should identify measurement requirements, for aerosol properties and distributions, for research that needs to be initiated with minimum delay.

This chapter is intended to provide an overview of effects on the Earth's radiative balance caused by changes in stratospheric and tropospheric aerosols (i.e., climate forcing by aerosols). After defining terminology, the nature of forcing by greenhouse gases is compared with that by aerosols. A description is presented of the evidence for climate forcing from secular increases in anthropogenic tropospheric aerosols. Whereas volcanic eruptions are decidedly nonanthropogenic, the aerosol they inject into the stratosphere affords an atmospheric experiment into the radiative effects of aerosols of unparalleled magnitude. Current knowledge about effects from episodic occurrences of stratospheric aerosol increases is then reviewed. Finally, the nature of climate forcing by different classes of anthropogenic aerosols is discussed.

AEROSOL RADIATIVE FORCING OF CLIMATE

Climate forcings are changes imposed on the planetary heat balance that alter global temperature (Hansen and Lacis, 1990; Hansen et al., 1993a). Although the term climate forcing is sometimes used to refer to changes in energy balance associated with internal fluctuations of the climate system (e.g., regional cloud or snow cover), we restrict the word *forcing* to describe anthropogenic or other *externally imposed* changes in energy balance. Such changes are measured in watts per square meter, and they allow direct comparison of forcing from different atmospheric constituents. The current, global mean forcing from anthropogenic increases in GHGs [including CO_2, CH_4, N_2O, and chlorofluorocarbon (CFC) increases since ca. 1800] is estimated to be approximately +2.5 W m^{-2} (IPCC, 1995a). Substantial geographical variability in this GHG forcing exists mainly as a result of differences in the Earth's surface temperature. By comparison (as discussed below), the stratospheric aerosol from the eruption of Mt. Pinatubo caused a temporary global mean climate forcing (by scattering sunlight back toward space) of about -4.0 W m² [i.e., exceeding GHG forcing in magnitude (opposite sign)]. Both models and measurements show a transient (1992-1994) response of about -0.5°C in global mean surface temperature attributed to the Mt. Pinatubo aerosol (McCormick et al., 1995). Although large geographical and temporal (diurnal to seasonal) variability existed from the volcanic aerosol, there was clear evidence of a transitory aerosol-induced effect on the radiative balance of the Earth resulting from the eruption of Mt. Pinatubo.

While in the atmosphere, aerosol particles affect the energy balance of the Earth both *directly* [by reflecting and absorbing shortwave (solar) radiation and by absorbing and emitting some longwave (infrared) radiation] and *indirectly* [by influencing the properties and processes of clouds and possibly, although these effects are unknown, by modifying atmospheric dynamics and chemistry (e.g., by participating in the heterogeneous chemistry of reactive greenhouse gases such as O_3)].

The *direct* effect can be observed visually as the sunlight reflected upward from haze when viewed from above (e.g., from a mountain or an airplane). The result of the process of *scattering* of sunlight is an increase in the amount of light reflected by the planet and, hence, a decrease in the amount of solar radiation reaching the ground. Possible meteorological consequences of the direct effect range from a global change in energy balance to changes in the rate of warming and drying of soil or the rate of evaporation of surface water.

The *indirect* aerosol forcing of climate is a result of the observation that anthropogenic emissions cause increases in the number of particles that can nucleate cloud droplets. As a consequence, the number concentration of cloud droplets, which is governed, in part, by the number concentration of aerosol particles in the pre-cloud, is also increased. An increased number concentration of cloud droplets leads, in turn, to enhanced multiple scattering of light within clouds and to an increase in the optical depth and albedo of the cloud. The areal extent of the cloud may also increase as a result of several different mechanisms (e.g., larger droplet numbers yielding smaller droplets, slower coalescence, and thus longer droplet lifetimes, or organic material slowing the rate of evaporation of droplets). A key measure of aerosol influences on cloud droplet number concentrations is the number concentration of cloud condensation nuclei (CCN). These are particles that will activate to form cloud droplets at a given supersaturation of water. Other meteorological influences, such as changes in precipitation, might occur as a result of perturbations in the number concentration of aerosols; however, such effects have yet to be assessed quantitatively. Even though these indirect effects are not as theoretically tractable as the direct ones, they may be as important as (or more important than) direct effects in contributing to radiative forcing. For example, calculations suggest that a global change of -/+15 percent in the droplet population of marine low (stratus) clouds would cause a change in energy balance of +/-1 W m^{-2} (Charlson et al., 1992a).

Table 1.1 compares the elements of climate forcing by aerosols with that for GHGs. Two major aspects of the comparison of anthropogenic GHG and aerosol forcing are noteworthy:

1. The two forcings (GHG and aerosol) have very different spatial and temporal distributions: GHG forcing operates night and day, whether clear or cloudy, and is at a maximum in the hottest, driest places on Earth (most infrared radiation is trapped in the atmosphere by vapor, liquid, and solid forms of water). In contrast, forcing by anthropogenic aerosol occurs mostly by day, is strongest without clouds, and because of the relatively short residence time of aerosols, is concentrated near aerosol sources.

2. Most GHGs have lifetimes in air very long compared with lifetimes of aerosols. For example, the current, enhanced level of CO_2 represents an accumulation of many decades of emission, whereas the current anthropogenic aerosol derives from emissions during only the previous few days.

EVIDENCE FOR RADIATIVE FORCING BY ANTHROPOGENIC AEROSOLS

The first question the panel addressed was: is anthropogenic aerosol forcing of climate a sufficiently important element of the overall climate system, such that its uncertainty is limiting our ability to quantitatively assess the effect of anthropogenic emissions on climate change? The panel finds that the evidence supporting an aerosol forcing effect on climate of a magnitude comparable, but opposite in sign, to that of GHGs over industrialized regions of the Northern Hemisphere is compelling. It is the opinion of the panel that the uncertainty in the magnitude of the effect of aerosols on climate is seriously hindering our ability to assess the effect of anthropogenic emissions on climate.

There are a number of independent lines of evidence supporting the hypothesis that anthropogenic aerosols cause substantial climate forcing that is of a magnitude comparable to that of GHGs, but opposite in sign. The following lists some of this evidence, subdivided into direct and indirect forcings.

Direct Forcing

1. Observations show that stratospheric aerosol from the 1991 eruption of Mt. Pinatubo produced a peak global mean optical depth at 550-nm wavelength of 0.1 to 0.2, which resulted in a measured peak global forcing

TABLE 1.1 Comparison of Climate Forcing by Aerosols with Forcing by Greenhouse Gases (GHGs): Fundamental Differences in Approach to Determination and Nature of Forcing

Factor	Long-Lived GHGs (CO_2, CH_4, CFCs)	Short-Lived GHGs (O_3, HCFCs, VOCs)	Aerosols
Optical properties	Infrared absorption is well quantified for all major and minor GHGs	Infrared absorption is reasonably well quantified	Refractive indices of pure substances are known, but size-dependent mixing of numerous species and the nature of mixing have optical effects difficult to quantify
Important electromagnetic spectrum	Almost entirely longwave ($\lambda > 1\mu m$)	For O_3, solar and longwave are important	For tropospheric aerosol, mainly solar; for stratospheric aerosol, solar and longwave contributions lead to stratospheric warming
Amounts of material	Well mixed; nearly uniform within the troposphere	Highly variable in space and time. Concentrations may be estimated by chemical models, but with considerable uncertainty	Pronounced spatial and temporal variations
Determination of forcing	Well-posed problem in radiative transfer; originally considered by Arrhenius (1896)	Radiative aspects well posed. Global networks provide some data to test model predictions of geographical and altitudinal distributions	*Direct*: Relatively well-posed problem, but dependent on empirical values for several key aerosol properties. Dependent on models for geographical/temporal variations of forcing. *Indirect*: Depends on aerosol number distribution; no fundamental approaches are yet available. Inadequacy of mathematical descriptions of aerosols and clouds seriously restricts abilities to predict indirect forcing

Dependence of forcing on loading (at present)	Varies as weak function (square root or logarithm) of concentration	Generally nonlinear with concentration; for halocarbons, a linear dependence	*Direct:* Almost linear in the concentration of particles *Indirect:* Undoubtedly nonlinear
Nature of forcing	Varies geographically, from ~0.6 W m^{-2} at South Pole to 3 W m^{-2} in the Sahara region. Forcing is exerted at the surface and in the troposphere. Operates night and day	Strongly dependent on geographical, altitudinal, and temporal variation (e.g., O$_3$). Forcing is exerted at the surface and troposphere. Operates night and day	Tropospheric forcing varies strongly with location and season and occurs only during daytime; maxima of forcing occur near sources and at the Earth's surface. Stratospheric forcing includes some longwave effect but is dominated by shortwave radiation (daytime only); following major volcanic events, stratospheric mixing yields a forcing that is substantially global in nature For nonabsorbing tropospheric aerosols, forcing is almost entirely at surface; for stratospheric aerosols, there is a small heating resulting in a transient warming of stratosphere

NOTE: CFC = chlorofluorocarbon; HCFC = hydrogenated chlorofluorocarbon; VOC = volatile organic compound.

of about -4.0 W m^{-2} in 1992 (McCormick et al., 1995) and a temporary, calculated and observed cooling of the surface of approximately 0.5°C (Lacis, 1995). Optical depths from anthropogenic aerosol in industrial regions of the Northern Hemisphere are often greater than 0.1 or 0.2 and cover large portions of the Northern Hemisphere.

2. Studies of visibility indicate increasing extinction from the 1940s to the 1970s as a result of anthropogenic aerosols in the eastern United States (Husar et al., 1981). Typical visually estimated extinction coefficients of (1-3) x 10^{-4} m^{-1} along with an estimated scale height of 2 km yield aerosol optical depths of 0.2-0.6 in agreement with observations. Regional-scale optical depth estimates coupled to a regional-scale atmospheric sulfur model for the eastern United States (Ball and Robinson, 1982) were suggested to produce an annual average loss of solar irradiance of 7.5 percent from sulfate and other anthropogenic aerosols relative to preindustrial times. There has been a corresponding, measured 3 to 4 percent loss of solar irradiance over Europe per decade during the past 40 years (Liepert et al., 1994).

3. Measured light scattering by atmospheric aerosols is highly correlated with the measured masses of sulfate and organic compounds in the aerosol. A multiple regression analysis for the United States yielded a squared correlation coefficient of 0.95 (White, 1990). Maps of SO_2 emission, sulfate aerosol concentration, acidic deposition, and extinction coefficient show geographical coherence over the eastern United States (Husar et al., 1981; Charlson et al., 1992b).

4. Advanced Very High Resolution Radiometer (AVHRR) satellite imagery shows an enhanced aerosol optical depth in the Northern versus the Southern Hemisphere, with maxima in the vicinity of industrial regions (Durkee et al., 1991). In addition, the Stratospheric Aerosol and Gas Experiment (SAGE) data show a spring- and summertime enhancement of aerosol extinction by about a factor of 3 in the Northern Hemisphere midlatitude upper troposphere versus the Southern Hemisphere (Kent et al., 1991). As noted earlier in conjunction with Figure 1.1, most anthropogenic sulfate is produced in the Northern Hemisphere.

5. Numerous quantitative estimates of the direct effect of anthropogenic sulfate alone have been made based on models of atmospheric radiative properties. Zero-, two-, and three-dimensional models of the sulfur cycle have been coupled to radiation transfer calculations, resulting in estimates from -0.3 to -1.3 W m^{-2} for the global average forcing. The two- and three-dimensional models suggest annual average maxima in the Northern Hemisphere midlatitudes from -4 to -11 W m^{-2}. Uncertainties are estimated crudely as a factor of 2. Direct forcing by aerosols from biomass combustion has an even greater uncertainty, ranging from -0.05 to -0.6 W m^{-2} globally averaged. The Intergovernmental Panel on Climate Change (IPCC) has agreed on a "best estimate" of -0.4 W m^{-2} for sulfate alone based on

Kiehl and Rodhe (1995), -0.2 W m^{-2} for the contribution from biomass burning and +0.1 W m^{-2} for soot based on Haywood and Shine (1995). A direct forcing of -0.5 W m^{-2} may have discernible climatic effects if it is regionally inhomogeneous. A more complete summary of direct forcing estimates is provided later in this report.

6. It is currently estimated that about half of the background optical depth is the result of anthropogenic sulfate and organic aerosols (Andreae, 1995). As a result, based on a calculated global-scale cooling of 2-3°C from background aerosol (Coakley et al., 1983), a 1-1.5°C global-mean cooling can be estimated to be caused by anthropogenic aerosol.

Indirect Forcing

To quantify indirect climatic effects of aerosols requires relating increased mass concentrations of aerosol from anthropogenic sources to increased number concentrations of aerosol particles, to increased numbers of cloud condensation nuclei (CCN), to increased numbers of cloud droplets, to altered cloud radiative properties or lifetime. Increased droplet populations cause increased albedo for fixed liquid water path; therefore, increased numbers lead to negative forcing. Few studies exist that demonstrate the cause-and-effect relationships among these factors, although there are considerable data relating to the effect of anthropogenic emissions on CCN concentrations or cloud properties:

1. It is well established that CCN concentrations are greater in anthropogenically influenced continental air masses than in the marine atmosphere; CCN concentrations in maritime air uninfluenced by anthropogenic emissions rarely exceed 100 per cubic centimeter (cm^3), whereas concentrations in well-aged continental air generally exceed 1000 cm^{-3} (Pruppacher and Klett, 1978). Twomey et al. (1978) reported measurements of CCN concentrations exceeding 4500 cm^{-3} after relatively clean air (CCN levels as low as 50 cm^{-3}) passed over an industrial area in southeastern Australia. Radke and Hobbs (1976) reported CCN concentrations of 1000 to 3500 cm^{-3} in air advecting off the eastern seaboard of the United States. Hudson (1991) reported CCN measurements along the west coast of the United States, indicating a background marine concentration of 20 to 40 cm^{-3}, nonurban inland concentrations in Oregon of 100 to 200 cm^{-3}, and urban concentrations in the vicinity of Santa Cruz, California, of 3000 to 5000 cm^{-3}. Frisbie and Hudson (1993) report CCN concentrations of 500 and 5000 cm^{-3}, respectively, upwind and downwind of Denver, Colorado.

2. It has long been recognized that continental clouds tend to exhibit greater cloud drop number concentrations (CDNCs) than do marine clouds (Pruppacher and Klett, 1978). Numerous studies exist that link high CDNCs

to identified industrial sources (Mészáros, 1992). Warner and Twomey (1967) found that the number concentration in clouds downwind of sugar cane fires averaged 510 cm^{-3}, compared with a concentration of 104 cm^{-3} in the upwind maritime air. Fitzgerald and Spyers-Duran (1973) found that clouds downwind of St. Louis, Missouri, were comprised of higher concentrations of smaller droplets than the upwind clouds. A similar study was conducted by Alkezweeny et al. (1993) on cloud droplet size distributions upwind and downwind of Denver, Colorado. They found that the droplet size distribution in the downwind air was shifted to smaller diameters than that of the nonurban air (median volume diameter of 14 versus 28 μm) and the droplet concentration was increased by an order of magnitude (22 versus 226 cm^{-3}). Hudson and Li (1995) found significantly higher droplet number concentrations, and smaller mean drop diameters, in "polluted" clouds compared to "nonpolluted" clouds. Drizzle appeared to be suppressed in the polluted clouds.

3. Cloud drop number concentrations tend to increase with increased aerosol loading. Pueschel et al. (1986) reported measurements of cloud and aerosol concentrations at Whiteface Mountain, New York. The CDNCs increased strongly with aerosol loading in air trajectories that could be traced back to large urban areas in Pennsylvania.[1]

4. "Ship tracks," linear features of high cloud reflectivity embedded in marine stratus clouds, apparently result from aerosols emitted or formed from the exhaust of ships' engines (Coakley et al., 1987; Scorer, 1987; Radke et al., 1989a; King et al., 1993). Aircraft observations have shown enhanced droplet concentrations and decreased drop sizes in the ship tracks themselves, compared with adjacent, unperturbed regions of the clouds (Radke et al., 1989; King et al., 1993).

5. A number of studies present data indicating the sensitivity of cloud droplet number concentration to increase in aerosol concentrations. These include Warner and Twomey (1967), Pueschel et al. (1986), Gillani et al.

---

[1]Enhancing the concentration of CCN does not always lead to higher cloud drop concentrations. Similarly, enhancing cloud drop concentrations does not always lead to increased liquid water contents and cloud albedos. For example, high industrial emissions of CCN can actually lead to increases in rainfall and subsequent cloud dispersion (cf. Hobbs et al., 1970; Mather, 1991). Recent analysis of such phenomena by Cooper et al. (1994) has suggested that quite modest increases in 0.5-μm diameter particles can significantly enhance precipitation. Similarly, recent modeling work has shown that increases in droplet concentration do not always lead to decreasing drizzle and increased liquid water even in marine stratiform clouds (Ackerman et al., 1995). For continental conditions, in which the susceptibility to albedo modification by increase in precursor CCN concentrations is, in any case, low (cf. Twomey, 1991), increasing anthropogenic emissions may or may not lead to increases in low cloud extent, or cloud albedo.

(1992a, b), Leaitch et al. (1992), Berresheim et al. (1993), Hegg et al. (1993), Quinn et al. (1993), Raga and Jonas (1993), Martin et al. (1994), and Novakov et al. (1994). Whereas there is substantial variation in the exact magnitude of the sensitivity of CDNC to aerosol loading among the various studies cited, all are broadly consistent with an increased cloud drop concentration with increased aerosol loading for clouds that are homogeneously mixed.

6. A firm theoretical basis exists relating cloud albedo to cloud drop size distribution. Whereas it has proved difficult to demonstrate the theory in the field because of the inherent variability of cloud albedo, there are a number of studies that illustrate the relation between cloud drop microphysical properties and cloud albedo. Durkee (1988) found an enhancement of reflectivity of marine stratus clouds influenced by the urban plume from San Francisco, from the 3.7-μm channel of the AVHRR. Kaufman and Nakajima (1993) related reflectance (as measured by AVHRR) of low cumulus and stratocumulus clouds over Brazil to enhanced aerosol loading from biomass burning. Grovenstein et al. (1994) presented measurements of cloud albedo, derived from the 0.63-μm channel of the AVHRR, at Mt. Mitchell, North Carolina, and found a relation between cloud top albedo and cloud drop concentration, measured in situ, that agrees well with theory. Boers et al. (1994) examined monthly mean CCN concentration in marine boundary layer air at Cape Grim, Tasmania, versus optical depth of low marine clouds derived from satellite data; a trend of increasing optical depth with increasing CCN was found. Kim and Cess (1993), in evaluating monthly averaged albedo for low-level marine clouds as obtained from the Earth Radiation Budget Experiment (ERBE) for regions of the North Pacific and North Atlantic versus regions of the Southern Hemisphere, found albedo enhancement in coastal areas influenced by anthropogenic emissions.

7. Han et al. (1994) examined the latitudinal dependence of drop sizes of warm clouds from AVHRR for four seasonal months for clouds above land and ocean. It was found that the effective cloud drop radius is systematically smaller for continental clouds as opposed to maritime clouds. In addition, Han et al. (1994) found, in both continental and maritime clouds, systematically smaller effective drop radius in Northern Hemisphere midlatitudes. It was suggested that because of the effect of the inherent variability of liquid water path on cloud albedo, cloud drop radius is a more sensitive indicator of anthropogenic influence than is cloud albedo. Boucher (1995) constrained a global climate model with the satellite retrievals of Han et al. and determined the climate forcing that can be inferred from the observed distribution of cloud droplet radii. Based on two sets of experiments, the study suggests that the indirect radiative forcing by anthropogenic aerosols could be -0.6 or -1.0 W m^{-2} averaged in the 0°-50°N latitude band. The uncertainty in these estimates was judged to be at least 50 percent.

8. An analysis by Hansen et al. (1995) suggests that observed changes in diurnal temperature range (DTR) are most likely caused by increases in low-level cloud cover over large regions of the Northern Hemisphere. Although these cloud changes are consistent with expected consequences of increasing concentrations of anthropogenic airborne particles, other explanations for these observations are possible.

9. The observed pattern of temperature change shows no increasing likeness to that predicted by climate models that do not include forcing by anthropogenic aerosols, whereas there is an increasing correlation with time between the observed pattern of temperature change and patterns predicted with climate models that do include anthropogenic sulfate aerosol forcing (Mitchell et al., 1995; Santer et al., 1995a). The 50-year trend in the pattern correlation statistic in the summer (June, July, August) and fall (September, October, November) seasons is statistically significant when measured against trends deduced from the same analysis applied to data derived from unforced climate model simulations. The predicted pattern of vertical temperature change is also consistent with the historical record when cooling by aerosols is included (Santer et al., 1995b). Although they are consistent with a sulfate effect on climate, it is not possible to unambiguously attribute the proposed trends to the specific mechanisms and forcings included in the models at this time.

These lines of evidence taken together, along with the relative weakness of evidence to the contrary, are convincing to us that climate forcing by anthropogenic aerosol is sufficiently significant to be considered as roughly equivalent (but of opposite sign) to that of GHG forcing in many parts of the world, although the uncertainty is large. In both cases—direct and indirect—the uncertainties are unacceptably large and present serious limitations to climate modeling.

We recommend that the uncertainties in calculated aerosol forcing at the top of the atmosphere be reduced to within ±15 percent both globally and locally. This limit of uncertainty is equivalent to that required in estimating the greenhouse gas forcing (IPCC, 1995a). Locally this would imply an uncertainty in forcing of less than 1.5 W m^{-2} (by assuming a local aerosol effect of -10 W m^{-2} in the diurnal mean).

EVIDENCE FOR CLIMATE RESPONSE TO ANTHROPOGENIC AEROSOL FORCING

A number of studies that either estimate a quantitative contribution to or suggest a role in radiative forcing by aerosols are mentioned above. Climate response, as evidenced, for example, in the temperature record, represents the entire climate system's integrated response to all forcings as well as natural variability. Detection of an unambiguous climatic response to any particular forcing mechanism, such as that resulting from aerosols, has proved to be extremely difficult. It can be asked: to what extent does the observed climate record reflect the effect of possible aerosol forcing? The answer will likely depend on what climate record is considered–namely, what part of the Earth, what period of time, and so forth. The possible options at this point are that an observed climate record is consistent with, inconsistent with, or shows no correlation with an aerosol forcing effect.

Ice core records from Greenland and Antarctica indicate that there have been enhancements in the concentrations of aerosols during different time periods (Legrand, 1995). For example, there have been strong increases in the concentrations of dust particles at the high latitudes during cold climates. Numerous increases in sulfur species also appear to have occurred for short time spans (<1-2 years) in the past following major volcanic eruptions. Since the beginning of this century, sulfate concentrations occurring in Greenland snow have been considerably enhanced, due to mostly increased fossil fuel combustion, whereas no similar trend is seen in Antarctica. Thus, the paleoclimate record from ice cores has proved to be an extremely useful indicator of the fluctuations occurring in aerosol concentrations at high latitudes, in both the near and the distant past. However, more research is required to translate unambiguously the aerosol record in the ice cores into a measure of the global atmospheric aerosol loadings prevailing during the different time periods, thus enabling quantitative assessments of their role in past climates.

Although a comparison of estimated global means might suggest that anthropogenic aerosol to some degree has compensated for GHG forcing, the spatial/temporal differences preclude any simple cancellation. Not only are GHG and aerosol forcings fundamentally different, so that their climate effects cannot be expected a priori to compensate for each other even if global mean values of radiative forcing cancel, but also fundamentally different approaches are required for determining their respective magnitudes. Further, because of the different atmospheric lifetimes, responses to increases or decreases in sources of the two classes of compounds will be different.

Diurnal, seasonal, and spatial differences in aerosol forcing provide opportunities to seek an aerosol fingerprint in the Earth's temperature record (Karl et al., 1995). As noted earlier, for example, the direct reduction in surface radiative flux from aerosols is limited to daylight hours, unlike GHG forcing. Because of the spatial heterogeneity of aerosols, as opposed to GHGs, the spatial heterogeneity of aerosol forcing may be more likely to be reflected in spatial differences in temperature records.

In recent decades, surface temperatures appear to have risen more in the Southern than in the Northern Hemisphere (Wigley, 1989), contrary to many model predictions for GHGs alone (e.g., Manabe and Wetherald, 1980). Engardt and Rodhe (1993) compared seasonal mean temperatures during two 20-year periods (1926-1945 versus 1971-1990). In areas where the mean annual column burden of sulfate exceeded 8 milligrams (mg) per square meter, summer temperatures averaged 0.4°C cooler than in regions where the reverse was true. Winter, on the other hand, was 0.5°C warmer in the high SO_x source regions. Regions of high sulfate concentrations warmed faster than low-sulfate areas in the summer between 1910 and 1940 but cooled more rapidly from 1950 to 1970. Whereas relative cooling in 1950-1970 during summer is consistent with an aerosol forcing, the temperature response for the 1910-1940 period and the winter warming are not. Trends in Northern Hemisphere temperatures (and cloudiness possibly associated with these temperatures) since about 1950 are proximately associated with changes in the large-scale dynamics of the atmosphere (Norris and Leovy, 1994; Wallace et al., 1995). Whether these changes are permanent is unknown, and the extent to which aerosol or other factors are responsible for them is also unknown. A decrease in the diurnal temperature range over industrial regions has been observed, which may be partially attributable to the presence of anthropogenic aerosol forcing (Karl et al., 1995). One concludes from available climatic data that some surface temperature and cloud cover observations are consistent with an effect of aerosol forcing whereas others are not. The statistical significance issue, particularly if natural variability is included, makes it impossible to come to concrete conclusions at this time.

Cooling of the Northern relative to the Southern Hemisphere occurs in numerical model simulations that incorporate tropospheric aerosol forcing (Taylor and Penner, 1994; Roeckner et al., 1995).

Model computations indicate that the transient increase in the stratospheric optical depth from the eruption of Mt. Pinatubo resulted in a temporary cooling of the surface by about 0.5°C (Lacis, 1995). This is in approximate agreement with observations (McCormick et al., 1995), thus indicating the potential for a cooling of the climate from increases in sulfate aerosols.

We caution that the usefulness of this qualitative agreement between models of climate response and models of forcing is strongly dependent on the degree to which both types of models are realistic. It is extremely important to quantify forcings and responses independently; hence, until the sensitivity of climate models is verified independently (i.e., the magnitude of the derivative of surface temperature with respect to forcing dT_s/dF is determined), we can only conclude that the above agreement or disagreement between climate and aerosol forcing models and temperature data cannot be used to quantify the magnitude of forcing by aerosols or disprove the hypothesis that aerosol forcing exists.

Thus, climate forcing by anthropogenic aerosols appears to have partially masked the greenhouse effect over some parts of the Earth, particularly the Northern Hemisphere midlatitudes, perhaps reducing the rate of change of global mean temperature, and thus rendering the detection of global climate change more difficult than had been originally expected. Quantifying the magnitude and characteristics of the forcing by anthropogenic aerosols is therefore a necessary step in understanding global climate change.

RADIATIVE FORCING OF CLIMATE BY STRATOSPHERIC AEROSOLS

The quantification of climate forcing by stratospheric aerosols, together with climate model calculations of response, provides a unique and independent method of testing climate models. It is important also to quantify possible influences of volcanic aerosols on clouds and cloud forcing in order to be able to separate natural (volcanic) from anthropogenic aerosol forcings.

Changes in concentrations of stratospheric aerosols, caused by intense, episodic volcanic eruptions, perturb the climate system by (1) warming the lower stratosphere through enhanced absorption of solar and longwave radiation, and (2) reducing the solar radiation reaching the surface-troposphere system through increased albedo. The result is a negative radiative forcing of the surface-troposphere system. Consistent with the residence times of aerosols in the stratosphere (about one to two years), the direct climate forcing by a single volcanic eruption typically continues for only a few years. Increases in stratospheric aerosol concentrations also have the potential (yet to be quantified) to influence cloud formation and maintenance processes in the upper troposphere, circulation in the lower stratosphere, and lower-stratospheric ozone concentrations (via heterogeneous chemical reactions on particle surfaces).

In contrast to the cumulative, long-term, climate forcing from a well-mixed unreactive GHG having a long residence time (e.g., CO_2), the stratospheric aerosol forcing from a single eruption is transient. Because of the climate system's inertia to such a transient forcing, only a small fraction of the possible equilibrium temperature change can be realized, unlike the case for long-lived GHGs.

There are important differences in the status of observations for episodic stratospheric aerosol increases versus secular increases in anthropogenic tropospheric aerosols. Because of the much longer residence time of particles at high altitudes, stratospheric aerosol increases typically become nearly global in extent (during the few years following an eruption) for tropical eruptions. Consequently, satellites and remote sensing techniques are capable of obtaining globally relevant information on stratospheric aerosols. Quantitative consistency between satellite and aircraft/ground-based retrievals of stratospheric aerosol properties has been achieved (e.g., Russell and McCormick, 1989), providing robust estimates of transient stratospheric aerosol climate forcings.

Observations following the major eruption of Mt. Pinatubo have offered a wealth of details regarding forcing and response characteristics associated with stratospheric aerosols (McCormick et al., 1995). Indeed, several aspects of present theoretical knowledge on the radiative and climatic effects of volcanic aerosols have been strengthened in the wake of these observations, as outlined below.

Radiative forcing by stratospheric aerosols is governed essentially by the column burden of the particles and their sizes. Surface-troposphere forcing by stratospheric aerosols is less sensitive to aerosol composition and location within the stratosphere, but warming within the lower stratosphere does depend on altitude (Pollack and Ackerman, 1983; WMO, 1989). For particle effective radii of approximately 2 μm or less (typical for volcanic sulfate), cooling of the surface-troposphere system occurs (Lacis et al., 1992), with the sensitivity of the net radiative forcing estimated to be -2.5 to -3 W m^{-2} for an increase in midvisible optical depth of 0.1 (Harshvardhan, 1979; Lacis et al., 1992). Initially inhomogeneous, the forcing evolves spatially and temporally, consistent with microphysical mechanisms governing aerosol formation and removal, stratosphere-troposphere exchange, and global-scale transport.

The eruption of Mt. Pinatubo in June 1991 yielded optical depth perturbations ranging from about 0.1 to 0.3, varying with location and time. The most optically thick portions of the aerosol were located between 20 and 25 km and were confined to 10°S-30°N during the early period (see *Geophysical Research Letters* 19, 149-218, 1992). Within two to three months, perturbed stratospheric optical depths were observed to at least 70°N, along with an enhancement in the Southern Hemisphere. Directly observed narrow and broadband total solar irradiance effects (IPCC, 1995a) indicate that reductions in surface solar flux ranged from <5 to 20 W m^{-2} in the diurnal mean, depending on prevailing aerosol optical depth. Satellite observations indicate a global mean decrease of about 5 W m^{-2} in the absorbed solar radiation in the period immediately following the eruption (IPCC, 1995a).

Model computations of radiative forcing by Pinatubo aerosols are in broad agreement with the observations mentioned above. In one model study (Hansen et al., 1992), a global mean midvisible optical depth of 0.1 ten months after the eruption and decaying exponentially with a one-year time constant was considered. The corresponding global mean forcing was estimated to have a maximum value of about -4 W m^{-2}. The volcanic aerosol forcing during the first two years after the eruption is estimated to be greater than or comparable to GHG forcing over the past century and is substantially greater than the current decadal increase in GHG forcing (about 0.4 W m^{-2} per decade).

Evidence for Climate Response to Stratospheric Aerosol Perturbation

Stratospheric aerosol perturbations by volcanic eruptions tend to cool the surface-troposphere system. Time series of anomalies from satellite microwave measurements reveal that there was a nearly global, but nonuniform, tropospheric cooling following the Pinatubo eruption (Dutton and Christy, 1992), coinciding with the reduction in solar radiation reaching the troposphere. Cooling during the summer of 1992 was greatest in continental interiors (Hansen et al., 1993a). Although there are difficulties in isolating signals from volcanic versus other causes (Robock and Mao, 1994), attempts to identify patterns of climate response from large volcanic eruptions, by removing influences of El Niño/Southern Oscillation (ENSO) events, have shown that cooling following an eruption can last up to about two years, with an amplitude of approximately 0.1-0.2°C when averaged over the six largest eruptions of the past century.

A general circulation model (GCM) investigation of the climatic impact from the 1991 Mt. Pinatubo eruption, using an initial optical depth of about twice El Chichon's and thereafter decaying with a one-year e-folding time (Hansen et al., 1992), predicted a transient cooling, maximum amplitude, and temporal evolution agreeing reasonably well with observations. The model-computed surface cooling ranged from 0.4 to 0.6°C, whereas the observed cooling estimated in 1992 (about one year after the eruption) ranged from 0.3 to 0.5°C. GCM investigations also revealed the possibility of dynamically induced responses in the tropospheric circulation patterns caused by volcanic perturbations (Graf et al., 1993).

Heating of the tropical lower stratosphere resulting from an enhancement of the aerosol layer by volcanic injections leads to a transient, local temperature increase. This has been observed following eruptions of Mt. Agung, El Chichon, and most recently Mt. Pinatubo (IPCC, 1995a). Besides increasing temperature, the heating by aerosols could lead to anomalous upward motion in the lower stratosphere. This, in turn, would lead to adiabatic cooling and reductions in ozone concentrations (Kinne et al., 1992).

Increases in stratospheric aerosols raise the potential for ozone destruction in the lower stratosphere via heterogeneous reactions on particles (Hoffman and Solomon, 1989). Observations in 1993-1994 after the Pinatubo eruption indicate that ozone values fell to unusually low levels. Changes in ozone concentrations are known to perturb the radiative balance (WMO, 1992).

Satellite observations (Minnis, 1994) suggest the potential for significant cloud modifications and hence an indirect radiative forcing from volcanic aerosols. Polarization lidar observations by Sassen (1992) in the northern midlatitudes following the Pinatubo eruption indicate that resulting changes in upper-tropospheric aerosols may have modified the microphysical and optical characteristics of upper-tropospheric clouds.

Inferences from Stratospheric Aerosol Research

Major volcanic eruptions provide a test of our ability to model climate change caused by transient, near-global, aerosol-induced perturbations. For the Pinatubo eruption, the fact that both computed forcing and modeled response of the climate system are in reasonable agreement with observations is encouraging. An important implication is that fundamental aspects of our knowledge on climate, at least for impacts from a moderately large and global aerosol forcing of limited duration (two to three years), appear to be sound.

It is important to realize, however, that this volcanic aerosol test is an inappropriate analogue for a tropospheric aerosol forcing with a secular trend spanning several decades (Ramaswamy et al., 1995). In particular, the Pinatubo "experiment" inadequately tests the role of the oceans in climate change and neglects land-sea contrasts, characteristic of forcing from tropospheric aerosols. In many respects, the episodic stratospheric aerosol forcing problem is far less complicated than the tropospheric aerosol climate problem. Nonetheless, the relatively better understanding of stratospheric aerosol climate effects does provide an extremely useful reference with which to compare and contrast forcings and responses from secular increases in tropospheric aerosols, since the optical depths in the two cases are comparable, albeit operating on different time and space scales.

CLIMATE FORCING BY KEY TROPOSPHERIC AEROSOL TYPES

Historically, many different classification schemes have been used to describe key aerosol types (e.g., see d'Almeida et al., 1991). An alternative to classifying aerosols by regional types, based on the notion of mass balances within global biogeochemical cycles, has been used increasingly for connecting atmospheric aerosol mass concentrations and burdens to the source strengths of aerosols or their gaseous precursors. Some of these aerosol types are defined simply by the chemical cycle of an element (sulfur, carbon, nitrogen, etc.), whereas others relate to a particular source (soil dust, sea salt, etc.). Because there is a wide variety of sources and molecular forms for carbonaceous aerosols, for example, there are several mass-balance cycles to consider for this type of aerosol (photochemical oxidation products of natural and anthropogenic hydrocarbons; smoke from biomass combustion; and soot, particularly from fossil fuel combustion).

Table 1.2 provides estimates of global source strengths and resultant optical depths for natural and anthropogenic aerosol types (Andreae, 1995). Whereas the anthropogenic aerosol mass flux is estimated to be only about 10 percent of the total, perhaps as much as about 50 percent of the global mean aerosol optical depth is anthropogenic; the reason follows from both shorter lifetimes and lower optical extinction efficiencies for soil dust and sea salt aerosols than for the major anthropogenic types (sulfates and smoke from biomass combustion). Because Table 1.2 displays the components of optical depth for each aerosol type, it is possible to estimate the increment of climate forcing from each type, as shown in Table 1.3. The estimated global mean optical depth figures are based on the "best-estimate" source strength given in Table 1.2, assumed average lifetimes, and mass extinction coefficients at ambient relative humidity for each of the aerosol types. No formal estimate of the overall uncertainty is given, nor is it available; however, despite the uncertainty that is implicitly present (e.g., in the range of flux estimates of Table 1.2), the singular message is clear. That is, given current best estimates of source strengths and aerosol properties, anthropogenic sulfates, organics, and soot are very likely to contribute a substantial fraction of the aerosol optical depth of the whole atmosphere.

Table 1.3 summarizes estimates of climate forcing for sulfates, smoke from biomass combustion, and soot, with an additional entry for chemically undifferentiated "anthropogenic aerosols." With the exception of the early, low estimate by Bolin and Charlson (1976), which was based on the then very uncertain assumption—now known to be wrong—that anthropogenic sulfates are present over only 1 percent of the globe, all of the estimates of global mean forcing by sulfate alone fall in the range from -0.3 to -1.3 W m^{-2}. The multiple-box models are based on only three chemical models (Langner and Rodhe, 1991; Pham, 1994; Taylor and Penner, 1994) and give global mean anthropogenic sulfate forcings ranging from -0.3 to -0.9 W m^{-2}. Analysis of uncertainties of the estimates in Table 1.3 has thus far been cursory, ranging from a stated factor of 2 (Charlson et al., 1991) to approaches using the square root of sum of squares (Charlson et al., 1992a; Penner et al., 1994b). Consequently, given inadequate uncertainty analyses along with incomplete sensitivity tests, a single best estimate (and its uncertainty) of the global mean forcing is currently unavailable. Comparison of the calculated magnitudes of forcing by individual components in Table 1.3 suggests that the magnitude of the net forcing could be small if the lower magnitudes of sulfate and organic aerosol forcing are combined with the higher values for soot. It is clear that the magnitudes of the uncertainties at present are too large to allow a resolution of this situation and that more measurements

TABLE 1.2 Source Strength, Atmospheric Burden, Extinction Efficiency, and Optical Depth for Various Types of Aerosols

Source	Flux (Tg/yr)			Mass Extinction Coefficient ($m^2 g^{-1}$)	Estimated Global Mean Optical Depth
	Low	High	Best		
Natural					
Primary					
Soil dust (mineral aerosol)	1,000	3,000	1,500	0.7[a]	0.023
Sea salt (mass mean diameter = 5 μm, σg = 2)	1,000	10,000	1,300	0.4[b]	0.003
Volcanic dust	4	10,000	33	2.0	0.001
Biological debris	26	80	50	2.0	0.002
Secondary					
Sulfates as $(NH_4)_2SO_4$ from natural precursors	85	210	102	5.1	0.014
Organic matter from biogenic VOCs (as C)	40	200	55	5.0	0.011
Nitrates from NOx	15	50	22	2.0	0.001
Subtotal	2,200	23,500	3,062		0.055
Anthropogenic					
Primary					
Industrial dust, etc.	40	130	100	2.0	0.004
Soot (elemental carbon)	5	20	10	10.0	0.003
Secondary					
Sulfates as $(NH_4)_2SO_4$ from SO_2	120	250	140	5.1	0.019
Biomass burning (as C)	50	150	80	5.0	0.017
Nitrates from NOx	25	65	36	2.0	0.002
Organic from anthropogenic VOCs (as C)	5	25	10	5.0	0.002
Subtotal	300	650	376		0.047
Total	2,500	24,000	3,438		0.102

NOTE: VOCs = volatile organic compounds.

[a]Another value for the soil mass extinction coefficient is 1.0 $m^2 g^{-1}$ (cf. Malm et al., 1994), and values well above 2 $m^2 g^{-1}$ have been inferred (e.g., Ouimette and Flagan, 1982). White has recommended a value around 0.7, that given in this table (White, 1990).

[b]The sea salt mass extinction coefficient is based on a purely coarse mode sea salt, but recent studies have suggested many more submicron particles than had previously been believed (cf. O'Dowd and Smith, 1993; McInnes et al., 1994). The mass scattering efficacy could be higher than the value of 0.4 $m^2 g^{-1}$ cited.

ADAPTED FROM: Andreae (1995); IPCC (1995); Kiehl and Rodhe (1995).

are needed to resolve the issue of the overall magnitude of anthropogenic aerosol forcing.

Similarly, predictions for geographical and seasonal variations of forcing by anthropogenic sulfates differ among available models. These variations may be more important than differences in global mean values for estimates of meteorological effects. Despite these differences in predictions, all models show large negative forcings over the industrial regions of the United States, Europe, and Asia that are greater than calculated GHG forcings in these regions. Because there have been no estimates of uncertainties of predicted geographical variations, selection of a best or most probable estimate is impossible. Nonetheless, data exist for industrial regions that are consistent with predictions of large negative forcings (Liepert et al., 1994); it is therefore unlikely that predictions are so inaccurate that sulfate aerosol forcing can be neglected. Uncertainties in direct forcings by aerosol types other than sulfate are even less reliable; therefore, the same general conclusion applies—that selection of a "best estimate" or "central value" is fraught with unacceptably large uncertainties.

For indirect forcing, the most obvious unifying theme among available predictions is author admission of model inadequacies. If the observed decreased droplet size in Northern Hemisphere clouds relative to those in the Southern Hemisphere given by Han et al. (1994) is caused by anthropogenic aerosol, the required approximately 10 to 50 percent estimated enhancement of CCN would correspond to forcings in the range of about -0.5 to -3 W m^2 (Kaufman et al., 1991; Boucher and Rodhe, 1994; Jones et al., 1994). This estimate, along with the frequently demonstrated observation that anthropogenic pollution enhances CCN population (thereby causing larger numbers of smaller droplets), strongly suggests that some amount of negative forcing exists and that its magnitude may be significant (e.g., see Hobbs et al., 1974).

Table 1.4 lists key anthropogenic aerosol types, their forcing mechanism(s), and brief assessments of current understanding.

CONCLUSIONS

It is our judgment that

- climate forcing by anthropogenic aerosols is likely to be of sufficient magnitude to necessitate its representation in models of climate change over the industrial period; and
- present estimates of anthropogenic aerosol forcing are sufficiently uncertain as to be inadequate to usefully represent this forcing in models of climate change over the industrial period.

TABLE 1.3 Estimates of Direct Climate Forcing ($W\ m^{-2}$) by Anthropogenic Aerosols

Aerosol Type	Global Mean Forcing	Regional Maximum Forcing	Reference as Basis of Estimate
Anthropogenic sulfate alone	-0.1 to -0.2	-10 to -20	Bolin and Charlson (1976);[a] turbidity data
	-1.6	—	Charlson et al. (1990); single-box model
	-0.6	-4 (eastern Mediterranean Sea)	Charlson et al. (1991); global 3-D chemical/radiative model
		-2 (eastern U.S.)	Langner and Rodhe (1991); slow oxidation rate case
	-1.3	—	Charlson et al. (1992a); single-box model
	-0.3	-4.2 (Europe)	Kiehl and Briegleb (1993); 3-D chemical model and GCM
		-3 (eastern U.S.)	Same as Kiehl and Rodhe (1995, Figure 6b); slow oxidation case of Langner and Rodhe (1991)
	-0.66	-11 (central Europe)	Kiehl and Rodhe (1995), Pham (1994); chemical model
		-5 (eastern U.S.)	
	-0.45 (est.)	-6 (est.)	Kiehl and Rodhe (1995); standard oxidation case of Langner and Rodhe (1991)
	-0.95	-4 (over Europe)	Taylor and Penner (1994)
		-3 (eastern U.S.)	
	-0.56 to -0.94	—	Penner (1995); single-box model
	-0.36 to -0.79	—	Haywood and Shine (1995)

Organic, carbonaceous aerosols	-0.1	—	Penner et al. (1992); biomass combustion
Soot	-0.5	—	Penner (1995); single-box model
	+0.35		Penner (1995)
	+0.05 to +0.27		Haywood and Shine (1995)
Chemically undifferentiated or mixed anthropogenic aerosols	-1.4 to -2		Coakley et al. (1983)
	ca. -0.25 (sulfate plus soot)	—	Hansen et al. (1993b)
	-0.5 (total, including biomass smoke)		
	ca. -2 (summer)	-7 (summer 50°N)	Grassl (1988); estimated, based on zonal-mean model
	ca. -0.5 (winter)	-1.4 (winter 30°N)	No soot
	ca. -0.7 (summer)	-2.9 (summer)	Grassl (1988); including 20% soot
	ca. -0.2 (winter)	-0.4 (winter)	
	-0.29	-3 (Europe)	Haywood and Shine (1995). Langner and Rodhe (1991); slow oxidation case external mixture (7.5% soot)
		-2 (eastern U.S.)	
	-0.15	-2.5 (Europe)	Haywood and Shine (1995); internal mixture
		-1.5 (eastern U.S.)	

NOTE: 3-D = three-dimensional; est. = estimated.

^{a}Forcing was extracted from the Bolin and Charlson (1976) estimate of optical depth for sulfate by applying a sensitivity of forcing to optical depth of 30 W m^{-2} per unit optical depth (Charlson et al., 1991). Forcing by soot is taken to be a positive number that reduces the magnitude of negative forcing from sulfate because of the presence of soot.

TABLE 1.4 Key Anthropogenic Aerosol Types, Associated Forcing Mechanisms, and Status of Understanding

Key Anthropogenic Aerosol Types	Forcing Mechanism(s)	Status of Understanding
1. Water-soluble inorganic species (e.g., sulfate, nitrate, ammonium) from atmospheric reactions of precursor gases (e.g., SO_2, NO_x, NH_3); key sources include combustion of fossil fuels and smelting of sulfide ores	a. Direct, clear-sky upscatter of solar radiation	i. Fundamental scattering theory is fully understood for spherical particles; well in hand for nonspherical ones ii. In situ and remote measurement methods are available for quantifying the direct optical effect and the size-resolved chemical composition. Satellite measurements currently are limited; promising possibilities exist for large improvements iii. Coupling of local radiative properties of aerosol to energy balance is well understood iv. Relationship of aerosol optical and chemical properties to relative humidity is known in principle for pure compounds; the role of mixing with organic species is yet to be explored v. Quantitative connection has been attained relating SO_2 source strength to geographically dependent sulfate aerosol concentration; uncertainty approximately a factor of 2 in the newest models. Less is known about other compounds
	b. Indirect effect of CCN on cloud albedo	i. Theory suggests increased CCN number concentration should increase cloud albedo, if all other factors are held constant ii. Empirical evidence suggests that anthropogenic aerosols increase cloud albedo iii. Relationship of aerosol mass concentration to CCN number concentration is practically an open question
	c. Indirect effects of CCN on cloud droplet lifetime	i. Problem almost entirely open

2. Condensed organic species from atmospheric chemical reactions of reactive organic gases or from smoke produced by biomass combustion	a	i. Same as 1.a.i-iii ii. Connection to source strength of reactive organic gases is expected from laboratory work but not yet quantified in the field iii. Relationship to relative humidity remains to be explored
	b	i. Same as 1.b.i-iii ii. Role of water-soluble and partly soluble organics modifying cloud nucleating properties is an almost open question
	c	i. Role of organics in modifying cloud droplet lifetime is an open question
3. C(0), elemental or black carbon soot from incomplete combustion (e.g., diesel fuel)	d. Absorption of solar radiation; change in vertical temperature profile	i. Fundamental theory of light absorption by aerosols is well established ii. Radiative transfer theory for absorption is also well established iii. Data on presence or amount of light-absorbing particles (especially elemental and black carbon) are sparse; in situ methods are available iv. Connection of amount of absorbing aerosol to source field is possible in principle but remains to be done
4. Mineral dust, windblown soil, and desert dust	a, b, c, d, and absorption/ emission of longwave (IR) radiation	i. Same as 1.a.i-iii ii. No estimates yet available for fraction of soil dust that is anthropogenic iii. Little information is available on cloud nucleating properties of soil dust. May be a problem of second-order importance because of large particle sizes and relatively small number concentrations

We recommend that the uncertainties in calculated aerosol forcing at the top of the atmosphere be reduced to within ±15 percent both globally and locally. This limit of uncertainty is equivalent to that required in estimating greenhouse gas forcing (IPCC, 1995a). Locally this would imply an uncertainty in forcing of less than 1.5 W m^{-2} (by assuming a local aerosol effect of -10 W m^{-2} in the diurnal mean).

This report presents our opinions about needed scientific studies (measurements, observations, model developments); the technologies required (satellites, computers, aircraft, instruments); the necessary resources; and an implementation plan for a U.S. multiagency program to answer these scientific questions and, thereby, to improve climate models. Chapters 3 and 4 contain our recommendations for implementing a focused, integrated research program. First, however, Chapter 2 describes needed research.

2

Elements of a Research Program for Aerosol Forcing of Climate

The scientific problems that need to be addressed to narrow uncertainties in aerosol radiative forcing of climate require a highly coordinated observational, laboratory, and modeling program. Figure 2.1 shows a heuristic diagram of the general components of an integrated aerosol-climate research program. The required research is composed of (1) observations, both short-term intensive observational campaigns and long-term systematic monitoring (both satellite and in situ); (2) process studies, both theoretical and laboratory-based; and (3) model development and evaluation, ranging from process models directed at a single phenomenon to global climate models. The arrows connecting the boxes in Figure 2.1 indicate how results from each component influence other components (i.e., the needed integration).

The ultimate goal of the program is to provide quantitatively accurate calculations of direct and indirect anthropogenic aerosol radiative forcings, as indicated in Chapter 1. Knowledge of these aerosol forcings is necessary both to understand past trends in climate (such as surface temperature records) and to permit accurate projections of future climate change for climate assessments. To accomplish this goal requires improved methods of representing aerosol radiative forcing in global climate models. The purpose of this chapter is to recommend the components of an aerosol research program required to achieve that goal.

The ultimate integrator of our understanding of the effect of aerosols on climate is the global climate model. For this reason, we begin the chapter with a discussion of research needs in global climate modeling for aerosols.

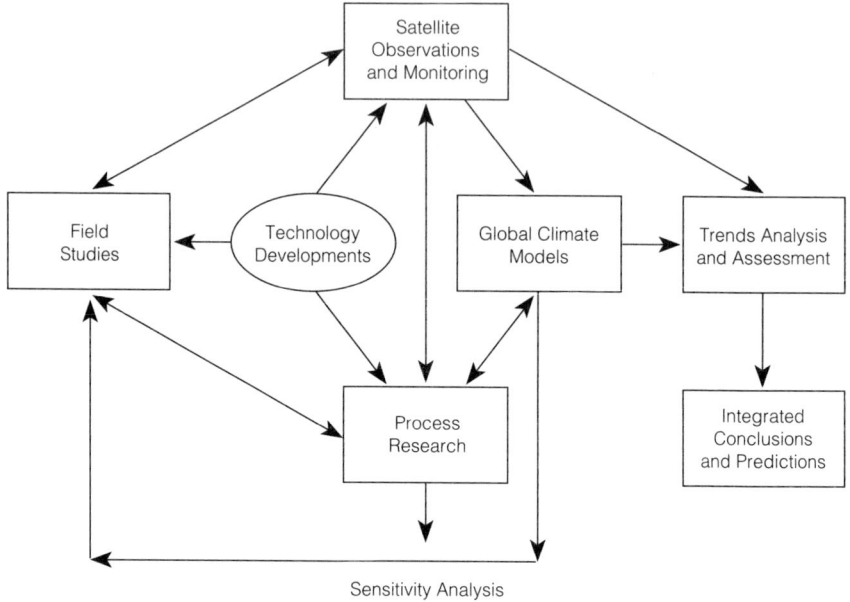

FIGURE 2.1 General components of an integrated aerosol-climate research program.

Aerosol processes that determine radiative forcing (direct and indirect) must be parameterized by way of *process models* that describe such phenomena as cloud formation and aerosol and cloud optical properties. Process models emerge from process research. Long-term systematic monitoring of aerosol properties can provide data to test not only process models but also global climate models. Intensive field campaigns provide an opportunity to measure a comprehensive set of chemical and physical parameters that govern features of aerosol forcing.

A method is required for determining both the individual phenomena that most influence climate forcing by aerosols and those that contribute the most uncertainty. As discussed in more detail in Chapter 3, sensitivity analysis (determining the change in model output resulting from change in each input parameter) is the traditional method. Figure 2.1 indicates that sensitivity analysis should ultimately be performed using both global climate and process models, since the magnitude of the sensitivity may depend on the scale of the phenomenon. These sensitivities are then used in principle to define future research directions to narrow the uncertainties associated with specific processes.

Since we are seeking to understand the impact of anthropogenic aerosols on climate, we should explain why we include research in the cleanest parts of the remote marine atmosphere. There are several reasons for this. First, in polluted air masses, the natural and man-made aerosols are so intimately mixed that the exact nature of the anthropogenic contribution becomes blurred. We cannot quantify man's contribution without thoroughly understanding the background aerosol. Second, in many cases, pollution aerosols blow offshore, mingle with marine aerosols, and are removed to the oceans. Yet we have an inadequate understanding of those marine loss processes to achieve the needed accuracy in our models. Finally, the simplicity of remote aerosol systems makes it possible to perform much more rigorous closure and process experiments than are possible in complex, spatially heterogeneous polluted air masses. Since the same basic principles will govern aerosol behavior in both environments, process studies in the simplest remote environments are likely to be more cost-efficient and productive than working in air masses where a wide range of sources might be responsible for observed changes over time and space. We must learn to walk before we can hope to run.

Ultimate success of the research plan advanced in this report rests on integration of the various components shown in Figure 2.1. The nature of this program as an integrated plan requires strong coordination. A structure to ensure this coordination is provided in Chapter 4. First, though, detailed information is given about the components shown in Figure 2.1.

GLOBAL CLIMATE MODELS

The geographic extremes in aerosol forcing are important to the response of the climate system to such forcing. Although globally averaged estimates of aerosol forcing may be reported, to properly construct a global average forcing requires a global-scale model.

A global climate model is the most comprehensive tool available for studying the importance of any forcing on the climate system, including aerosol. Such a model consists of a number of components: an atmospheric general circulation model (AGCM), some form of an ocean model, a sea ice model, and a land surface model. An atmospheric chemical transport model (ACTM) is used to simulate the distribution of aerosol. These models continue to grow in complexity as more detailed chemistry and removal processes are added. The AGCM uses information on aerosols to calculate radiative forcing and supplies information (e.g., winds) to the chemical model to determine aerosol distribution.

Calculation of direct radiative forcing may not necessarily require all processes described in an AGCM. Indeed, by its very definition, radiative forcing means that no atmospheric process has altered the climate state used to calculate the change in radiative flux caused by aerosol. Aerosol forcing calculations are typically performed in an AGCM, but employ only the radiative transfer part of the AGCM. There are now a number of global model studies of direct aerosol forcing for sulfate aerosols (Charlson et al., 1991; Kiehl and Briegleb, 1993; Taylor and Penner, 1994; Boucher and Anderson, 1995). Kiehl and Rodhe (1995) have reviewed several of these calculations and find roughly a factor of 2 difference in forcing, caused by chemical modeling differences. Calculations of the indirect effect (Chuang et al., 1994; Jones et al., 1994; Boucher and Lohmann, 1995) indicate a forcing magnitude similar to that of the direct effect.

Atmospheric General Circulation Models

Atmospheric general circulation models are numerical representations of the equations of motion and physical processes that define the workings of the atmosphere. These models are global in spatial extent and represent the equations of motion in either a physical grid space or spectral space (i.e., the dynamical fields are represented by a series of spherical harmonics). Climate models, by their very nature, require time integrations from one season to multiple decades. The horizontal resolution of the AGCMs is usually determined by computational resources. During the past decade there has been increased emphasis on higher spatial resolution. Current AGCMs employ resolutions from $8° \times 10°$ down to $1° \times 1°$. The number of vertical levels in these models ranges from around 10 to 20.

Physical processes included in the AGCMs are radiation, convection, and processes in the boundary layer and the surface layer. Cloud processes related to cloud amount and optical properties are typically included as a part of the radiation calculations. All of these physical processes operate on scales smaller than those resolvable in the AGCM and hence require parameterization. Much of AGCM development over the past 30 years has focused on improving these physical parameterizations, with much of the focus on clouds and convection. At present, there is a diverse range of approaches to the parameterization of these processes, which no doubt reflects limitations in understanding. A list, by no means complete, of some global models currently being used to study aerosol forcing is given in Table 2.1, where the atmospheric general circulation and chemical models are listed separately. The chemical transport models listed in Table 2.1B cover hemispheric to global scales and are suitable for coupling to AGCMs.

Atmospheric Chemical Transport Models

Predictions of aerosol distributions by an ACTM require information on magnitudes and geographic distributions of emissions of precursor gases and sources of primary particles, chemical reaction rates in the atmosphere, transport of these gases and aerosols by large-scale advection and subgrid-

TABLE 2.1 Global Models Currently Used to Study Aerosol Forcing

Model[a]	Horizontal Resolution	Number of Vertical Levels	Temporal Resolution (minutes)
A. Atmospheric General Circulation Models for Aerosol Forcing Calculations			
GISS	4° × 5°	9	7.5
LMD		11	6
MPI[b]	2.9° × 2.9°	19	24
NCAR[b]	2.9° × 2.9°	18	20
UKMO	2.5° × 3.75°	19	30
LLNL-CCM1	4.5° × 7.5°	12	30
B. Global and Synoptic Models for Chemical Transport of Aerosols			
LLNL Grantour	4.5° × 7.5°	12	6
Stockholm/Mainz Moguntia	10° × 10°	10	Monthly mean
NCAR IMAGES	2.5° × 2.5°	25	Monthly mean
GChM	1.0° × 1.0°	15	6
GISS	8.0° × 10°	9	6
MPI	2.9° × 2.9°	19	24

[a]GISS (Goddard Institute for Space Studies), LMD (Laboratoire de Météorologie Dynamique), MPI (Max Planck Institut für Meteorologie), NCAR (National Center for Atmospheric Research), UKMO (United Kingdom Meteorological Office), LLNL (Lawrence Livermore National Laboratory), Stockholm (Langner and Rodhe, 1991), IMAGES (Intermediate Model for the Annual and Global Evolution of Species), GChM (Global Chemistry Model; Pacific Northwest Laboratory).

[b]The MPI and NCAR models are spectral; the horizontal resolution listed is an equivalent Gaussian grid.

scale convection, and removal mechanisms (both wet and dry). Links to an AGCM are contained in large-scale advection of gases and aerosols, which employs atmospheric winds, and in convection parameterization, which supplies convective fluxes for subgrid vertical transport. Links are also associated with the land and surface modules in climate models through the representation of dry deposition processes. Wet removal processes in the ACTM require precipitation information from the AGCM. Finally, aqueous-phase reactions require information on cloud water. AGCMs are just beginning to prognostically calculate amounts and distribution of cloud water.

The development of many of the current ACTMs began with models that were not coupled to AGCMs. This uncoupling was necessary for accurate testing of transport and chemistry algorithms. These "off-line" ACTMs employ either AGCM atmospheric state information (e.g., winds, precipitation, cloud variables) or meteorological or climatological state information.

A number of the off-line models used for predicting sulfate aerosol distributions employ climatological data. Some questions remain about how accurate aerosol distributions are when calculated by using climatological data, since temporal changes in aerosol distribution occur on shorter scales than climatological and the resulting average aerosol distribution may not be well represented from climatological data.

To date, much of the global ACTM modeling has focused on sulfate aerosols (Langner and Rodhe, 1991; Chuang et al., 1994; Pham, 1994). These models have treated the aqueous and gas-phase production rates of sulfate in a highly simplified manner, for example, by assuming that the rate of aqueous formation of sulfate is proportional to cloud amount and cloud lifetime. This assumption does not properly represent the rate of formation of SO_4^{2-} in clouds when H_2O_2 is the limiting species (nor are the roles of trace metals and other oxidizing agents accounted for) and therefore does not represent the expected seasonal variation of sulfate aerosol. Other models that attempt to account for oxidant limitation in simplified ways predict the seasonal cycle of SO_4^{2-} better in source areas over North America than in Europe (Penner et al., 1994a). Reasons for lack of observed seasonal variation in sulfate over Europe are not understood. Oxidation of SO_2 on sea salt particles is another process not presently represented in ACTMs. This process operates mainly in the marine boundary layer, and the amount of sulfate produced in this manner is, as yet, poorly quantified. Yet it may be important to represent this process in global aerosol models, to obtain a more complete understanding of the atmospheric sulfur cycle.

Only a few model calculations are available of global-scale distributions and climate forcing by components of the aerosol other than sulfate, and these have been limited to smoke aerosols from biomass burning and/or black carbon from fossil fuel combustion (Penner et al., 1993; Cooke and Wilson, 1995; Liousse et al., 1995). This latter component is especially important for determining the light absorption coefficient and single-scatter albedo (a measure of the relative magnitudes of aerosol scattering and absorption) of anthropogenic aerosols. As yet, no global-scale model calculation of the combined effects of all anthropogenic aerosol components (sulfate, nitrate, ammonium, organic carbon, black carbon, anthropogenic dust aerosol) and their climate forcing is available, although one-box model-based analysis of all these components has been carried out (Pilinis et al., 1995).

Additionally, to test model predictions against measurements (thereby performing a "closure" experiment), it is necessary that natural components as well as anthropogenic components be represented. Major uncertainties surround the prediction of aerosol components other than sulfate, with the primary uncertainty related to inadequate specification of source rates. In the case of carbonaceous aerosols (organic and black carbon), little information regarding natural and anthropogenic source strengths is currently available (Penner, 1995).

Existing models for aerosol forcing calculations do not explicitly carry detailed information on particle size; the approach is to predict aerosol mass. It is well established, however, that both direct and indirect radiative effects of aerosols can depend strongly on the size distribution of the aerosol (Boucher and Anderson, 1995; Nemesure et al., 1995; Pilinis et al., 1995). Thus, global aerosol models will eventually need to include explicit aerosol size information.

Attempts to calculate indirect climate forcing by anthropogenic aerosols have all been based on predicted geographical distributions of anthropogenic sulfate aerosol (Boucher and Rodhe, 1994; Chuang et al., 1994; Jones et al., 1994; Boucher and Lohmann, 1995). These approaches empirically relate cloud drop number concentrations, and thus cloud optical properties, to sulfate aerosol mass concentration derived from a chemical transport model. Whereas this approach is arguably the best starting point to evaluate the indirect effect, the underlying assumptions do not address physical cause-and-effect relationships in a fundamental manner. For example, emissions of aerosol precursors and primary emissions are known to vary by region. Such variations will alter the ratio of SO_4^{2-} to other aerosol components by region. Also, the functional relationship between cloud condensation nuclei (CCN) concentration and sulfate mass is expected to be nonlinear. For example, if most anthropogenic sulfate is formed via aqueous processes in clouds, then adding more SO_2 will mainly form larger particles rather than additional CCN, and only a small fraction of anthropogenic sulfur emissions would form new CCN (e.g., see Leaitch et al., 1992), although the climate forcing might still be significant (Chuang and Penner, 1995). Most of these models have further assumed that observed relationships between either cloud water sulfate and cloud droplet concentrations or between cloud base aerosol concentrations and cloud droplet number concentrations could be used to represent the response of cloud droplets to new aerosol particles or sulfate mass. It is not clear that these relationships can be used to analyze the response of climate forcing to either increases or decreases in anthropogenic aerosol emissions, because such changes may involve a change in aerosol size distribution. However, one attempt to account for the effects of changes in size distributions, which notes the importance of the aqueous pathway, still finds significant forcing (Chuang et al., 1994). Data relating sulfate and/or other aerosols to cloud droplet concentrations are sparse to nonexistent for remote marine clouds and can give conflicting low susceptibility when account is taken of the type of mixing processes within the clouds (Novakov et al., 1994). These processes have not been accounted for in any model estimates of indirect climate forcing.

Because prediction of both homogeneous and heterogeneous rates of formation of sulfate depends on understanding and predicting the photochemistry of ozone and other oxidants, especially H_2O_2, understanding sources of discrepancy between modeled abundances and observations may well require development of coupled global aerosol-photochemical models. Photochemical models of the global troposphere are just now becoming available (Crutzen and Zimmerman, 1991; Atherton, 1994; Müller and Brasseur, 1995), and none has been self-consistently integrated with an aerosol model, even in simplified form.

Finally, the coupling of the atmospheric and chemical components into a complete system model will no doubt introduce new nonlinear interactions within the coupled model. For example, addition of aerosols to the system will have a tendency to lower surface temperatures. This reduction in surface temperature can, in turn, lead to changes in evaporative fluxes, which can alter the hydrologic cycle (e.g., reduction in precipitation). Changes in the hydrologic cycle will then modify chemical production and removal of aerosols. Each link in this feedback loop is highly nonlinear and can alter the sensitivity of the overall climate system. The addition of feedback processes involving atmospheric aerosols is imperative for a credible climate system modeling program and represents a major challenge for atmospheric science in the twenty-first century.

Recommended Research on Global Climate Modeling of Aerosol Radiative Forcing

Global climate modeling of aerosol radiative forcing requires the following:

1. major advances in the representation of indirect climatic effects of aerosols in global climate models by treating in a fundamental manner the relationship between aerosol mass and aerosol number, aerosol number and CCN number, CCN number and cloud drop number concentration (CDNC), and CDNC and cloud optical properties;

2. development and evaluation of coupled global aerosol-photochemical models;

3. coupling of atmospheric chemical transport and aerosol models into a global climate model system;

4. improvements in models of the sulfur cycle and development of models for other aerosol types and for mixtures—it is particularly necessary to develop models with a time resolution that is short compared to those of the relevant meteorological processes (e.g., synoptic time scale); and

5. evaluation and representation of aerosol sources and precursor gases for aerosol chemical models.

PROCESS RESEARCH

Process studies are an essential part of the overall program design, because they provide the means to identify, understand, and quantitatively describe the processes controlling aerosol forcing of climate, as well as to develop and test parameterizations used in models, and because existing knowledge of the pertinent processes is inadequate to the required level of certainty. Evaluation of the sensitivity of process model results to uncertainties in model formulation and input parameters is necessary for guiding observational programs. Some process studies are appropriately conducted in the laboratory, and some require small-scale field experiments, whereas others involve major, multi-investigator, and multiplatform field campaigns. Because of the significant costs associated with large field campaigns, such studies are treated separately from smaller-scale process research.

For both direct and indirect forcing of climate by aerosols, the connection between properties such as aerosol mass and the actual radiative forcing is mediated by a series of intervening linkages, some of which are poorly characterized. For one family of important species, sulfates, these linkages are shown schematically in Figure 2.2. Process studies seek to elucidate the individual linkages, some of which are very complex and nonlinear, such as that between sulfate aerosol mass and CCN concentration, as noted earlier. Although process studies are generally focused on a single linkage or process, they often benefit from the context provided by conducting them alongside related observations in an integrated field program.

As noted earlier, many of the properties, both physical and chemical, that determine aerosol forcing result from processes that occur on scales unresolved by general circulation models (GCMs). These processes therefore need to be parameterized in terms of prognosticated large-scale variables (e.g., temperature, specific humidity, cloud water). A complicating factor in these parameterizations is that many of the processes are highly nonlinear, for example, the relationship between changes in SO_2 emissions and CDNCs. To develop and test these parameterizations requires process models. For example, fine-scale cloud models (two or three dimensional) that explicitly resolve scales of motion down to 500 to 1000 m, with explicit treatment of cloud microphysics, have been used to study the effect of increased CCN on cloud droplet number (Flossmann and Pruppacher, 1988).

We begin this section with a summary of optical properties, then proceed to aerosol formation, transformation, and removal processes, and finally consider aerosol process models that integrate all of these features.

Optical Properties

The basic quantities needed to describe the direct interaction of aerosol particles with solar radiation are the aerosol optical depth $\delta(\lambda)$, single scat-

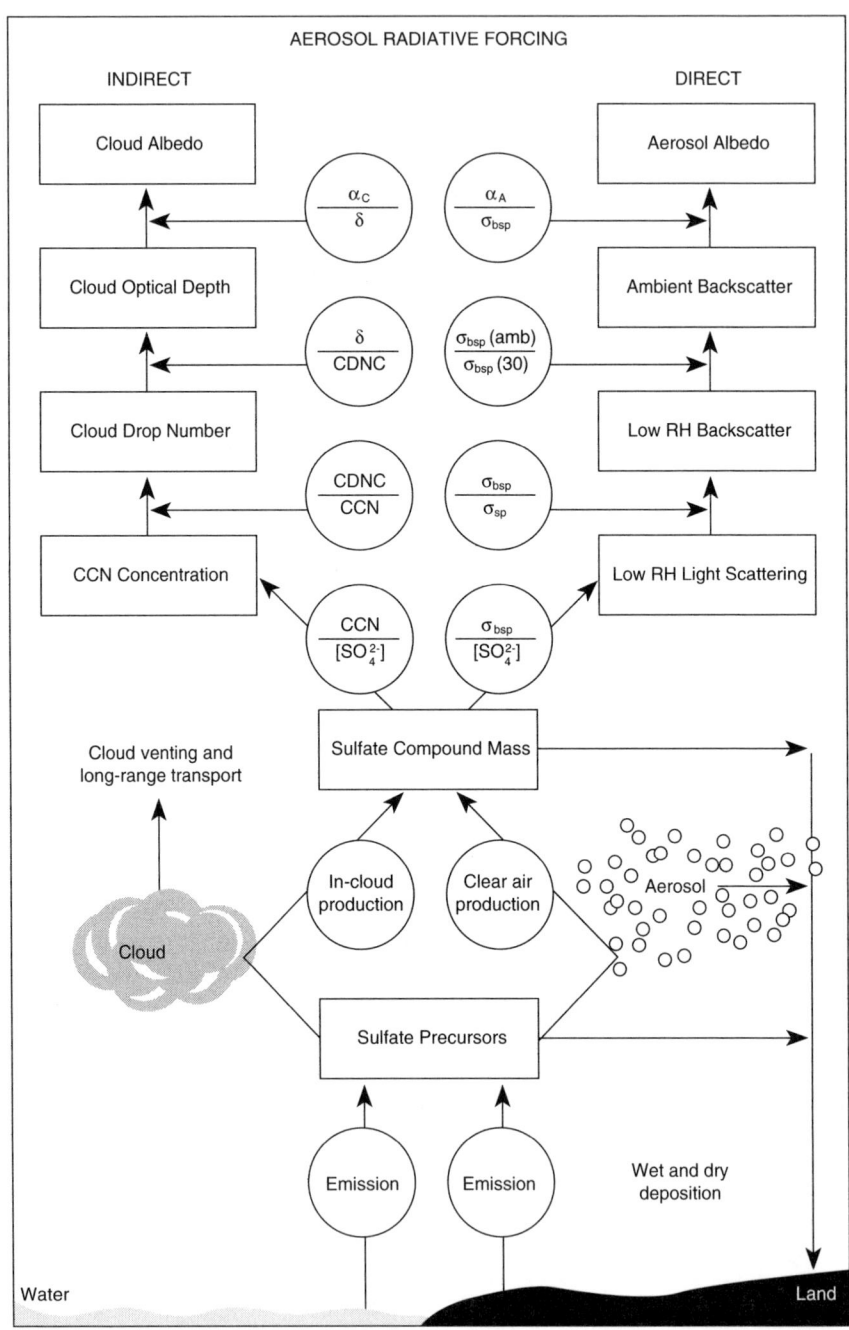

FIGURE 2.2 Direct and indirect forcing mechanisms associated with sulfate aerosols.

tering albedo $\omega(\lambda)$, and angular scattering function $\beta^*(\phi)$. The aerosol optical depth is the vertical integral of the aerosol extinction coefficient $\sigma_e(\lambda) = \sigma_{sp}(\lambda) + \sigma_{ap}(\lambda)$. The single scattering albedo, a measure of the relative magnitudes of aerosol scattering and absorption, is defined as $\omega = \sigma_{sp}/\sigma_e$. It can be measured directly in situ by separate measurement of σ_{sp} (with an integrating nephelometer) and σ_e (with a long-path extinction cell or atmospheric path) or from separate determinations of σ_{sp} and σ_{ap}. The angular scattering function describes the angular distribution of the intensity of light scattered by particles. Parameterizations in radiative transfer models use an integral property of the angular scattering function, either the asymmetry factor g, defined as

$$g(\lambda) = \frac{\int \beta^*(\phi,\lambda)\cos\phi\, d\cos\phi}{\int \beta^*(\phi,\lambda)\, d\cos\phi},$$

or the upscatter fraction $\beta(\mu)$, defined as the fraction of solar radiation scattered in the upward direction (back to space), where μ is the cosine of the solar zenith angle (Wiscombe and Grams, 1976). The angular scattering function $\beta^*(\phi)$ can be determined with a polar nephelometer (Hayasaka et al., 1992; Jones et al., 1994), but commercial versions of this instrument are not available. No instruments exist that directly determine g or $\beta(\mu)$. However, an integrating nephelometer with a special backscatter shutter can determine the hemispheric backscattering fraction $b = \beta(1)$, which can be used to estimate the value of the asymmetry factor.

The notion of describing sulfate forcing separately results in a negative forcing term ($\delta F/\delta SO_4^{2-}$), whereas $\delta F/\delta$ soot is positive (see Table 1.3). Absorbing aerosols can have either a positive or a negative radiative forcing (i.e., produce either a heating or a cooling effect), depending on the surface albedo and the ratio of aerosol light absorption to hemispheric backscattering coefficients $(1-\omega)/b$ (Chylek and Coakley, 1974; Haywood and Shine, 1995). For a given surface albedo, there exists a critical value of $(1-\omega)/b$ below which the aerosol leads to cooling, and above which to heating. The few measurements that are available (Waggoner et al., 1981) yielded values of $(1-\omega)/b$ in the range 1-3 in rural areas and 3-9 in urban areas. For the radiative transfer model used by Chylek and Coakley (1974), these results would imply a positive forcing in rural areas with surface albedos larger than 0.13-0.27. This range of surface albedos includes values expected for rural areas and points out the need for systematic, combined measurements of aerosol single scattering albedo and hemispheric backscatter fraction in order to represent properly the radiative effects of absorbing aerosols. Liousse et al. (1995) deduce global maps of ω that show, in general, a cooling effect of aerosols.

Given particle size distribution and chemical composition, Mie theory permits calculations of aerosol optical properties (Bohren and Huffman, 1983), but a number of key assumptions are required. Calculation of optical properties usually assumes that particles are spherical. This assumption is likely to be acceptable for hygroscopic particles, but may be poor for hydrophobic aerosols (e.g., elemental carbon and dust aerosols). Also, the state of mixture of different chemical species must be known. For internally mixed particles, which contain two or more chemical species, "mixing rules" for calculating the mixture's refractive index exist but must be evaluated, particularly for the case where insoluble, light-absorbing material is mixed with soluble, nonabsorbing species. In all cases, the wavelength-dependent refractive index of the major chemical species must be known. From a three-dimensional modeling perspective the important questions are how to combine (or mix) various aerosol optical properties on a GCM grid (around 300 km) and the degree of variability of these optical properties in both space and time, which is linked mainly to variabilities in aerosol properties, especially size distribution.

There is a clear need for laboratory determinations of the refractive index and density of pure and mixed-composition aerosol particles as a function of relative humidity. Models that relate the size and optical properties of particles to their chemical composition rely heavily on thermodynamic data for extremely concentrated solutions. Such data are lacking for certain components of sea salt and most organic species, even though excellent electrodynamic balance methods now exist for conducting such studies. In addition, there is a clear need to understand the relationship between various measurements of black carbon and the absorbing component of aerosols, which vary depending on the measurement methods being used and the source of black carbon in the aerosol sample (Novakov and Corrigan, 1995).

Two types of *closure* experiments of aerosol optical properties are needed (see discussion of closure experiments below). Local closure experiments test the suitability of Mie theory for predicting aerosol light scattering and absorption coefficients from measured size distributions and chemical composition. Such experiments are particularly important for aerosols that do not adhere to the assumptions of sphericity or homogeneity used in Mie calculations. Column closure experiments are also needed to test the consistency of radiative perturbation calculations using measured aerosol optical properties with their measured perturbation of radiative fluxes.

Relationships between cloud drop size distribution and cloud optical depth, on the one hand, and optical depth and cloud albedo, on the other, have been formulated based on approximate models whose validity is not well characterized. Comparison between measurements of cloud drop size distribution as a function of altitude within clouds and directly measured cloud optical depth are necessary to test various parameterizations currently used in models. Similarly, relationships between optical depth and cloud albedo should be tested, particularly as a function of spatial scale. Even if the extant parameterizations hold at particular points, it is unclear that this implies agreement on larger scales. Such problems as cloud edge effects, fractional cloudiness, and general inhomogeneity could invalidate larger-scale satellite retrieval without the employment of sophisticated algorithms that have yet to be formulated. In general, more comparisons of remotely retrieved cloud albedos with in situ measurement are in order.

Comparison of remotely retrieved and locally measured cloud albedos involves a difficult scaling issue. Cloud optical depths derived from in situ measurements are generally at a finer resolution than for remotely retrieved albedo (at least for currently available sensors). The issues of fractional cloudiness, edge effects, and cloud inhomogeneity complicate comparison between scales. Airborne sensors at various altitudes above cloud top would provide a useful link between in situ measurements and satellite observations, for which atmospheric corrections will complicate the scale problem.

Recommended Process Research on Aerosol and Cloud Optical Properties

The following research on aerosol and cloud optical properties is needed:

1. laboratory and theoretical determinations of the refractive indices of pure and mixed-composition aerosol particles, compared with bulk (e.g., filter samples) and in situ measurements of size-resolved composition and refractive index of atmospheric aerosols, to determine the accuracy of current theoretical treatments of refractive index of mixed-composition aerosols for use in radiative forcing calculations;

2. local closure experiments to evaluate Mie theory predictions of extinction with those measured;

3. comparison of theoretically calculated cloud optical depth and cloud albedo as a function of cloud drop size distribution with those measured as a function of cloud spatial scale; and

4. elucidation of theoretical issues relating remotely sensed albedo to that calculated based on in situ measurements of cloud drop size distribution.

Aerosol Dynamics

The size distribution and chemical composition of the atmospheric aerosol are determined by a number of physicochemical processes, including formation of condensable vapors, nucleation of aerosol particle embryos from these vapors, condensational and/or coagulation growth of the aerosol, and cloud processing. Elements necessary to understanding these processes are listed below.

1. *Identification of Aerosol Molecular Composition and Gaseous Precursors:* Much of the anthropogenic aerosol is secondary in nature, being derived through gas-to-particle conversion processes. Whereas the sulfate component of the tropospheric aerosol is well recognized, much less is known about the abundance and, in the case of organics, molecular composition of the other components of ambient aerosol. Accurate calculation of radiative and cloud nucleating properties requires knowledge of aerosol molecular composition, and aerosol molecular composition needs to be related to its gaseous precursors to be able to assess the affect of anthropogenic emission changes on aerosol climate effects. Little information on molecular speciation of the organic portion of ambient aerosols is currently available, precluding a proper linkage between organic gaseous precursors and organic aerosol components. Once identified, the optical properties of these components can be addressed (see previous section).

2. *Mechanism for Gas-to-Particle Conversion*: After low-volatility gases are generated by gas-phase chemical reaction, they may homogeneously nucleate or condense on existing particles. The dominating route determines the resulting aerosol number concentration and size distribution (Warren and Seinfeld, 1985). The rate and location of new particle formation by nucleation in the atmosphere are subjects of intense current debate (Clarke, 1993; Weber et al., 1995b; Raes and Van Dingenen, 1995; Russell et al., 1994). It is important to understand the competition that exists for gas-to-particle conversion in shaping the atmospheric aerosol size distribution since the aerosol number concentration depends crucially on this competition. Parameters that determine whether a gaseous species nucleates to form new particles or condense onto existing particles include the saturation vapor pressure of the condensing species [e.g., H_2SO_4, methanesulfonic acid (MSA), various organics, and NH_3], its accommodation coefficient for incorporation into existing particles, relative humidity, temperature, and the existing particle number size distribution.

Process studies that identify atmospheric conditions favorable for nucleation are crucial, since the generation of new particles is, at the same time, one of the most important processes controlling aerosol number concentrations and one of the most difficult processes to model. Short time-resolved measurements (order of a few seconds) of particle size distributions between 3- and 10-nm diameter can be used to identify events of new particle production in both the marine boundary layer and the free troposphere. These measurements, along with data for gas-phase aerosol precursors, permit computations of aerosol nucleation and growth rates. The actual nucleation of aerosol from precursor gases must be understood, theoretically, in laboratory studies and in the atmosphere. This understanding should be gained not just for single-component systems but also for multicomponent systems (i.e., more than one condensing species), such as those that actually exist in the atmosphere (e.g., H_2SO_4-H_2O-NH_3 and organics). Once again, much work is needed on organics. Theoretical formulations in process models need to be compared with laboratory and field results.

3. *Particle Growth Rates:* Particle growth occurs by condensation of vapors and coagulation of particles; as noted above, the available condensable vapors will be partitioned between the formation of new embryos and condensation onto preexisting aerosol. Understanding the rate of growth of particles is important in determining the factors that control particle size in regimes of effective light scattering and CCN activation. The time for growth of condensation nuclei to sizes where they may serve as CCN is important to particle residence time in the atmosphere. Competition between coagulation and condensation (for gas-phase production of condensable vapors) versus aqueous production of aerosol species determines the nature of the growing size distribution. Growth rates estimated theoretically and simulated in laboratory chamber experiments can be compared with measurements of the evolution of the atmospheric size distribution. Field experiments at several scales can also address this question. In a Lagrangian parcel experiment, an airborne platform (e.g., blimp or aircraft) follows an air parcel and periodically measures the aerosol size distribution. Another option is to sample at various points in a well-defined and steady-state circulation such as, for example, a land sea breeze circulation or, on a much larger scale, the Hadley cell circulation. The latter circulation has in fact been proposed as an incubator for the formation of CCN (Clarke, 1993).

4. *Hygroscopic Growth:* Optical and cloud nucleating properties of aerosol particles depend critically on their hygroscopic properties (Kim et al., 1993a,b; Tang and Munkelwitz, 1993, 1994a,b). Direct aerosol forcing sensitivity studies show that, of the variables on which aerosol forcing depends, relative humidity is the most influential (Boucher and Anderson, 1995; Nemesure et al., 1995; Pilinis et al., 1995). Although hygroscopic growth is involved in growth processes discussed in the previous section, the ubiquity of water in the atmosphere, its tremendous variability, and the existence of phenomena such as deliquescence and crystallization that have peculiar and large impacts on aerosol size all recommend treating, in a fundamental manner, the response of aerosol to change in atmospheric water vapor. Hygroscopic growth and shrinkage can be broken down into three facets. The first is a change in the physical size of aerosol particles as relative humidity changes, the second involves a change in aerosol index of refraction as water content of the aerosol changes, and the third is associated with a shift in particle size relative to the Mie scattering efficiency curve. All three effects combine to produce a change in radiative forcing of a given aerosol as a function of relative humidity. The

effects can be studied either individually (e.g., changes in aerosol size distributions with increasing relative humidity) or together. The latter type of study has been carried out, for example, by using two nephelometers sampling in parallel, the first at a reference relative humidity (e.g., 30 percent) at which the aerosol is essentially dry and the second with a variable humidifier to permit scanning the relative humidity from the reference humidity to 80 or 90 percent (Charlson, 1974). Hence, normalized scattering as a function of relative humidity is measured directly.

Deliquescence properties of ambient aerosols need to be measured and compared with those both measured in the laboratory for well-defined test aerosols and predicted based on thermodynamic theory. The extent to which ambient aerosols consist of "hygroscopic" and "nonhygroscopic" fractions in different regions, such as marine and continental, needs to be evaluated, and the dependence of these hygroscopic properties on chemical composition needs to be determined. Ultimately this information will serve as input to global aerosol models that have prescribed relative humidity fields and to GCMs that predict relative humidity fields.

5. *In-Cloud Processing*: Aerosol particles at a certain supersaturation of water can be activated to form cloud drops. Such drops can act as microscopic aqueous-phase reactors to produce sulfate from gaseous precursors and, as a result, yield still larger particles upon evaporation. By this route, CCN active only at relatively high supersaturations, such as those characteristic of cumulus clouds, can be transformed into CCN active at the lower supersaturations characteristic of climatologically important stratiform clouds. Furthermore, particles that are commonly too small to be optically active can be grown by this in-cloud mechanism into the optically active size range. Cloud processing of aerosol particles has been suggested to play an influential role in the marine boundary layer (Hoppel et al., 1994; Russell et al., 1995) and in the urban and regional atmosphere (Pandis et al., 1990; Meng and Seinfeld, 1994).

An important process study would involve simultaneous measurement of aerosol size distribution, CCN spectra, and chemical composition of the aerosol as a function of size. By using Köhler theory (Pruppacher and Klett, 1978), it should be possible to derive the CCN activation spectrum from size and composition data. A comparison between derived and measured spectra would constitute a needed closure experiment on Köhler theory. The comparison would also provide needed information on the fraction of CCN composed of particular chemical species (although some assumptions will be necessary) and, thus, on the sources of CCN. This study should address the impact of surface-active or sparingly soluble organic compounds on the activation of soluble particles: It has long been speculated that hydrophobic organic films on particles could inhibit their ability to nucleate cloud drops. The proposed comparative analysis could test this speculation, and if discrepancies between calculated and directly measured CCN activation spectra arise, laboratory studies should investigate plausible atmospheric surfactants that could be producing attenuation of CCN activity.

CCN activation spectra and cloud drop number concentrations are linked by supersaturation fields, which in turn are influenced by CCN spectra themselves as well as cloud dynamics. For an adiabatic cloud, the relationship is fairly straightforward, but most atmospheric clouds are not adiabatic, and therefore the functional dependence of the CDNC on the CCN activation spectrum is unclear, particularly near cloud top. A number of studies over the years have provided some support for a linear relationship, but the data are far from extensive. For clouds that undergo substantial entrainment, cloud dynamics could strongly modulate CDNC, thereby weakening the relationship between CCN and CDNC established at cloud base (Novakov et al., 1994). Measurements are needed of CDNC and interstitial CCN spectra as a function of altitude within clouds, coupled with CCN spectral measurements in entrained air.

Recommended Process Research on Aerosol Dynamics

The following research on aerosol dynamics is needed:

1. Determine the molecular composition, particularly of the organic fraction, of ambient aerosols and the relation of that organic fraction to gaseous precursors.

2. Develop an understanding of the atmospheric conditions that favor homogeneous nucleation of vapors to form particles for single and multicomponent systems, especially for sulfates, ammonia, and organics. Develop instrumentation for rapid measurement of ultrafine particles ($\leq$10-nm diameter), and deploy this instrumentation in regions likely to be sites of new particle formation such as the marine boundary layer and the free troposphere. Compare observed rates of new particle formation with those predicted by theory, where theory is available. Where it is not available, develop an appropriate theoretical framework for atmospheric nucleation.

3. Measure evolution of the atmospheric aerosol size distribution under well-defined conditions that allow one to assess the extent to which theoretical growth rate predictions conform with observations. Couple gas-phase and aerosol composition measurements to growth rate measurements.

4. Measure deliquescence properties of ambient aerosols and compare them with those both measured in the laboratory for well-defined test aerosols and predicted based on thermodynamic theory. The extent to which ambient aerosols consist of hygroscopic and non-hygroscopic fractions in different regions, such as marine and continental, needs to be evaluated, and

the dependence of these hygroscopic properties on chemical composition needs to be determined.

5. Perform simultaneous measurements of aerosol size distributions, CCN spectra, and aerosol composition as a function of size both in the laboratory for well-defined aerosols and in ambient air, and evaluate the extent of agreement with Köhler theory. Especially measure the organic fraction of the aerosol and its effect on activation.

6. Link subcloud CCN/CDNC/cloud albedo.

7. Develop and evaluate parameterizations to characterize and determine the importance of aqueous production of aerosol components in global models.

Aerosol Sinks

In dealing with atmospheric chemical transport models above, we noted that a primary uncertainty is related to the inadequate specification of source rates. Table 1.2 provided a qualitative estimate of these uncertainties, showing that many source terms are currently only "order-of-magnitude" estimates. In the immediately preceding section dealing with aerosol process models, specific questions have been raised about parameterizations for wet and dry removal processes. The goal of the present section is to outline research on sources and sinks needed to improve models of both direct and indirect radiative forcing by particles.

A central point in this section and in our recommendations for future research is the following: In essentially all previous research on aerosol sources and sinks, whether dealing with specific biogeochemical cycles or with specific applied problems (e.g., associated with atmospheric releases of radioactivity), the primary focus has been on aerosol (or chemical or radioactive) *mass,* which for spherical particles is proportional to the *third* moment of the particle size distribution. As a result, large particles were of dominant concern, and measurements and theory emphasized these large particles. As examples, substantial research has investigated the dependence on wind speed of airborne sea salt *mass*; many "acid rain" monitoring networks have provided data on sulfate (and nitrate) *mass* scavenged by precipitation; and most available data for dry deposition rates are heavily skewed by the deposition of the most massive particles. In contrast, for aerosol radiative forcing, lower moments of the particle size distribution are of dominant importance.

Thus, for the direct radiative forcing problem, of prime interest (to first order) is aerosol surface area, which for spherical particles is proportional to the *second* moment of the particle size distribution. Consequently, available data for both sources and sinks are, in many cases, of limited value to the direct radiative forcing problem, because previous data and analysis were "skewed" by previous emphasis on aerosol mass—current parameterizations of aerosol optical properties in terms of aerosol mass notwithstanding. Moreover, this limitation of previous data is even more apparent for the indirect forcing problem. In this case, the *number* of particles capable of acting as cloud condensation nuclei (at a specific supersaturation) or ice-freezing nuclei (IFN, at a specific temperature) is of prime interest; that is, of primary interest is the *zeroth* moment of the particle size distribution (with the integration starting from some minimum size for activating cloud particles under specific conditions).

Some limitations of information on particle mass alone can be illustrated with recently reported data obtained from sampling air near the Azores Islands. Garrett and Hobbs (1995) report that the number concentration of condensation nuclei (CN) increased by a factor of about 5 from a case of "clean maritime air" to one of "continentally influenced air," while the mass concentration of sulfate particles increased by a factor of about 100. In addition, for a different case study in the same region, Hudson and Li (1995) report that the concentration of CN (a measure of the total number of particles present) also increased by a factor of about 5 for "clean" versus "polluted" air, while the number of CCN active at 0.1 percent (largest particles) increased by a factor of about 50. Given these order-of-magnitude differences in results for "number" versus "mass," the emphasis of future research (for both sources and sinks) must be on particle size distributions, namely, the number of particles of each size class (and chemical, supersaturation, or temperature class, as appropriate), in contrast to the emphasis of previous research on particle mass.

Besides illustrating the importance of focusing on aerosol number rather than merely on mass, the data outlined in the previous paragraph can also illustrate types of research on aerosol sources and sinks that must be pursued to elucidate aerosol radiative forcing. In particular, if one tries to understand these data, layers of research questions about both aerosol sources and sinks appear. Thus, a first layer is revealed if one asks if these results mean that CN concentrations differ little in continental versus marine air (or polluted versus clean air) because the major difference in the aerosol content of these air masses is in mass loadings as a result of relatively rapid removal of CN (e.g., by coagulation); at this level of inquiry, a host of related questions can occur (e.g., dealing with coagulation rates, rates of removal of larger particles). A second layer is revealed if one asks if the results mean that our concepts of continental/polluted versus marine/clean are inadequate (e.g., because all marine air has been influenced earlier by continents) and that the governing feature is the time since the air mass experienced significant cleansing (e.g., by passing through a storm), significantly reducing aerosol mass but not CN; at this level, other related research questions arise (e.g., relative scavenging rates for larger sulfate particles versus CN). Still another layer is revealed if one asks if the dominant distinction in aerosol loading arises from natural production of CN (e.g., from natural sulfur sources) and if results such as these reveal mostly that air masses typically quite rapidly attain an "equilibrium" CN concentration, dictated by natural sources; again, a host of obvious research questions appear at this level (e.g., dealing with gas-to-particle conversion). Currently, such questions as these have not been answered adequately; it is therefore clear both that such questions must be addressed if aerosol radiative forcing uncertainties are to be reduced and that to answer such questions it is essential not only to obtain particle size information but also to examine sources and sinks simultaneously.

As already emphasized, reliable descriptions of aerosol radiative forcing require reliable numerical models of spatial distributions of size- and chemical-specific concentrations of aerosols. In turn, reliable models of these concentrations require reliable descriptions of wet and dry deposition. For the development of these descriptions, two concepts are critically important.

One of these concepts is that both the interactions of aerosols with radiation and aerosol removal processes are strongly dependent on particle size, especially in the particle-diameter range from about 0.1 to 10 μm. Over this range, wet and dry removal rates can vary by two to three orders of magnitude. Consequently, even if computational economy restricts radiative flux calculations in climate models to only a few bands of radiation, corresponding spectral average influences of aerosols require estimates of particle size-dependent aerosol concentrations, which in turn require estimates of particle size-dependent removal rates.

Another critically important concept is that macroscale consequences follow from microscale causes. Thus, atmospheric removal rates and therefore global-scale spatial distributions of aerosols depend on the abilities of submicrometer aerosol particles to act as CCN or IFN, the efficiencies with which particles are collected by vegetation elements such as pine needles and leaf hairs, dissolution in the sea of microscopic bubbles of air containing particles, etc. Consequently, for applications at macroscales, understanding at microscales is required.

If these two concepts are used to filter current knowledge about wet and dry removal processes, distressingly little of value for the radiative forcing problem is retained. It is true that, as a result of decades of international research on bomb debris fallout and acid rain, much is already known about wet and dry removal of aerosol particles. Illustrations of progress can be found in the three-volume proceedings of the Fifth International Conference on Precipitation Scavenging and Atmosphere-Surface Exchange Processes (Schwartz and Slinn, 1992), as well as in the proceedings of the four previous conferences in the same series; in acid deposition progress reports (e.g., from the U.S. National Acid Precipitation Assessment Program); and in the open literature. Essentially all of this information, however, has (appropriately) emphasized the deposition of particle mass, and therefore, most of the data are of little value for defining and helping to understand particle deposition as a function of particle size.

For example, throughout the world, many networks were established to monitor acid deposition, providing data on major ions in precipitation and some measures of dry deposition. Most of these data sets, however, are of little value for radiative forcing problems because of the data's (desired) emphasis on deposited mass. In this regard, notice that the mass of a single 1-µm (spherical) particle scavenged by precipitation is the same as the total mass of a thousand 0.1-µm particles (that might act as CCN), and that the mass of a single 10-µm particle (e.g., locally resuspended and then dry deposited) is the same as the total mass of a million 0.1-µm particles. Essentially all data for deposited radioactivity (e.g., Chamberlain, 1991) are of similarly restricted value for radiative forcing problems (direct and indirect) because, for bomb debris and nuclear accidents such as Chernobyl (e.g., Cambray et al., 1987), the radioactivity of each particle was proportional primarily to particle mass [although there were exceptions in which the radioactivity of each particle was proportional to its surface area (e.g., for radioactive species such as iodine that condensed on ambient particles)]. Consequently, there are few to no network data available to define wet and dry deposition of particles as a function of their size, which is critical for the radiative forcing problems.

Correspondingly, most of the available theoretical, semiempirical, and statistical models for wet and dry deposition of particles as a function of their size (and of a host of other variables, depending on the type of collector—from raindrops and ice crystals to forests and lakes)—have not been adequately tested against field data because such data are unavailable. In the few exceptional cases available, order-of-magnitude discrepancies exist, both between theory and data (e.g., for an important case in rain scavenging of particles; see Radke et al., 1992) and between data sets obtained by different experimental techniques (e.g., for an important case in dry deposition of particles to grass; see Garland and Cox, 1982). Consequently, additional field data to test and, as appropriate, revise available models of wet and dry deposition are essential to remove existing order-of-magnitude uncertainties in wet and dry removal rates, especially for particles in the size range 0.1-1.0 µm.

For the dry deposition of 0.1- to 1.0-µm particles, there are uncertainties of at least an order of magnitude even for their deposition to simple vegetative canopies (e.g., see Allen et al., 1991); greater uncertainties exist for forest canopies (e.g., see Peters and Eiden, 1992). For 0.1- to 1.0-µm particles depositing at sea, no particle size-dependent data appear to be available (e.g., see Rojas et al., 1993). Numerical studies have examined problems with modifying available dry deposition formulations for use in global models (e.g., Giorgi, 1988). In reality, the lack of mesoscale, particle size-specific field studies for dry deposition to inhomogeneous vegetation and to the ocean results in, at best, only order-of-magnitude estimates of dry deposition of particles at the global scale.

To obtain field data to test and improve models of dry deposition of particles, improved techniques and new instrumentation appear to be critical (Nicholson, 1988). Particle size-specific, eddy-flux measurements of ambient particles appear promising (e.g., Neumann and den Hartog, 1985) but to date have suffered from poor statistics associated with few particles in specific size classes. Further, similar studies at sea would have to account for sea salt particle production and for shifts in particle sizes from water vapor condensation. Also, preliminary measurements of size distributions of ambient particles within forest canopies have been valuable for estimating dry deposition to forests, but complications from emissions within the forest and mixing from above the canopy have not been adequately addressed. For size-specific dry deposition of particles in the radiatively important range of about 0.1 to 1.0 µm, further investigations should be undertaken of releasing essentially monodisperse particles that are tagged and then collecting and counting the number of tagged particles actually deposited. Such studies may be useful even at sea if the released particles can be, at once, hygroscopic or at least wettable, insoluble, and buoyant (e.g., monodisperse polystyrene microspheres coated with ammonium sulfate).

Turning now to wet deposition or precipitation scavenging, at the outset we want to emphasize two critically important and relatively new concepts dealing with vertical diffusion (and therefore long-range transport) of all chemicals (e.g., sulfur dioxide) that are important for modeling aerosol radiative forcing. In earlier studies, many predictions suggested that concentrations of relatively short-lived species such as SO_2 (typically oxidized in the atmosphere within a few days) would decrease relatively rapidly with height because, on average, vertical mixing in the troposphere proceeds with an average time scale of about a week. This average, however, is a spatial average and mostly reflects the relatively sparse spatial distribution of clouds and storms that have substantial updrafts. One important concept that apparently was not appreciated earlier was that, within these updrafts, vertical transport of even short-lived species could be substantial (Gidel, 1983). As a result, earlier estimates of concentrations in the upper troposphere of short-lived species were typically underestimated substantially, sometimes by many orders of magnitude.

The second, related and important concept is that these underestimates depend on the chemical species because vertical "diffusion" (or transport) in the atmosphere is species dependent. This dependence is a function not only of the lifetime of the species against oxidation (or other destruction), with mature convective storms able to transport species to the upper troposphere even if their lifetimes are only about 10^3 seconds, but also of the efficiency with which each species is scavenged by each storm. For example, a greater fraction of sulfur dioxide than sulfate mass entering a storm will usually be vented by it, because particles are scavenged by precipitation typically much more efficiently than SO_2 (unless the H_2O_2 molar concentration in the entering air is comparable to that of SO_2).

Currently, essentially all global-scale atmospheric chemistry numerical models contain parameterizations for this "sub-grid scale vertical transport" (e.g., Müller and Brasseur, 1995), but it is clear that much additional work is needed to improve these parameterizations (e.g., Lin et al., 1994). Some progress has been made in developing a "venting climatology" for different storms (e.g., Thompson et al., 1994) and, in testing venting parameterizations against predictions of cloud models (e.g., Pickering et al., 1995), but not only is there need to substantially increase the data base for transport by storms, there is also extremely limited information on the scavenging of different species, especially particles, by these same storms.

Very few studies of this type have been performed. In fact, the results outlined below are from the only two studies of storm venting of particles of which we are aware. By yielding conflicting results, these two studies illustrate the crude state of current knowledge about these critically important topics.

From upper-tropospheric measurements in cloud-free air recently mixed by convective storms in the midwestern United States, Kleinman and Daum (1991) deduced that only a few percent of the 0.1- to 1.0-μm particles in boundary-layer air ingested by storms survived transport of the air (and its insoluble constituents) to the upper troposphere. There are, however, several difficulties in interpreting the results of such clear air studies (e.g., accounting for other sources of insoluble gases such as CO, O_3, and NO_x; identifying regions of cloud outflow; and accounting for subsidence of outflow ice crystals and their particle loads, which would be expected to be different from the trajectory of insoluble gases). Nonetheless, additional field studies of this type would be useful.

A more focused study by Knollenberg et al. (1993) used the National Aeronautics and Space Administration's (NASA's) ER-2 aircraft to sample aerosols in anvils of convective storms at altitudes of about 15 km. Their results demonstrate that, with the associated cold temperatures, *mature cyclones* in tropical regions fully involve most condensation nuclei (CN, not just CCN) in ice nucleation processes. In contrast, for *isolated* anvils in the tropics and for anvil cirrus in continental regions, only a small fraction of total CN was found to be involved in ice nucleation: in the anvil of a continental cumulonimbus (over Arizona), CN concentrations continued to be in excess of 10^4 cm^{-3}. Additional studies, in which inert tracer gases and complete particle size distributions (and CCN and IFN spectra) are measured in both inflow and outflow regions of a variety of storms, would be extremely useful, especially if water fluxes were also defined so that scavenging efficiencies (or inefficiencies) could be related to precipitation efficiencies of the storms.

Also, we should mention other field studies, some supporting and some conflicting with earlier results, that have investigated a potentially important nonlinearity in scavenging of particles. Thus, Leaitch et al. (1992) report data confirming observations by Squires during the 1950s of cloud droplet concentrations increasing with CCN concentrations (see Twomey, 1977) and supporting Twomey's suggestion that increased concentrations of anthropogenic aerosols ingested by clouds could lead to increased cloud albedo (and increased cloud lifetimes). For warm stratiform clouds, Gillani et al. (1995) confirmed the findings of Leaitch et al. (1986) that the number of cloud droplets fails to increase linearly with the number of approximately 0.1- to 1.0-μm aerosol particles: cloud drop concentrations begin to saturate at about 500 cm^{-3}. However, other data (e.g., see Twomey, 1977, p. 177) fail to display this nonlinearity, even for droplet concentrations greater than about 1200 cm^{-3}. Additional studies are clearly needed, in which measurements should be made of not only particle size distributions but also of CCN spectra.

Finally, we want to emphasize the concept that understanding at the microscale is needed for applications at macroscales. For example, from measurements in both large and small capping cumulus clouds that ingest smoke plumes from biomass burns, Radke et al. (1992) not only found the expected rapid incorporation in cloud water of particles greater than about 0.4 μm in diameter (large enough to act as CCN in the smoke plume) but, for fires with large capping cumuli, found efficient removal of approximately 0.1-μm particles (with a minimum in the removal of particles in the intermediate size range from about 0.1 to 0.3 μm). Unfortunately, speculations about possible causes of this rapid scavenging of particles of approximately 0.1 μm are currently unconstrained by relevant data (e.g., for electrical charges in the more intense updrafts).

Recommended Process Research on Aerosol Sinks

To improve modeling of aerosol radiative forcing, additional and more complete wet and dry deposition field studies are especially required. For the needed dry deposition studies, which must focus on particle size-specific data, developments in technology (e.g., for eddy-flux measurements) and techniques (e.g., for measuring monodisperse particles actually deposited) appear to be necessary. The needed precipitation scavenging field studies should include tests of mesoscale models of precipitation formation, efficiencies, and scavenging. Especially necessary are aerosol particle size measurements of storm venting for a greater variety of pollution loadings and of cloud and storm types. Also, for both wet and dry deposition, measurements are needed of microscale quantities (e.g., electrical charges on cloud particles and pine needles, characteristics of dendrites on ice crystals and hairs on leaves) because only by understanding and accounting for the microscale processes that govern scavenging can we confidently develop reliable parameterizations for applications at larger scales.

The prime goal of the dry deposition studies is to obtain reliable, particle size-specific data in the field (i.e., not just from wind tunnels) for dry deposition to "real-world" collectors, from forests in inhomogeneous terrain to the oceans under a variety of conditions. To achieve the essential goal of obtaining particle size-specific dry deposition velocities, developments in technology (e.g., for eddy-flux measurements) and techniques (e.g., for measuring monodisperse particles actually deposited) almost certainly will be necessary. The prime emphasis of the precipitation scavenging field studies will be to develop parameterizations for storm venting for use in global-scale models for all relevant species (especially for particles as a function of their sizes), for all climatologically important cloud and storm types, and for representative ranges of pollution loadings and storm microphysical and dynamical variables. Analysis of field data must be performed with appropriate mesoscale models of precipitation formation, efficiencies, and scavenging.

Aerosols and Ice Formation in Clouds

Formation of the ice phase in atmospheric clouds is a strong modulator of the impact of clouds on global climate. This modulation arises principally in two fashions. First, cirrus clouds play an important role in the Earth's radiation balance, and the radiative properties of these clouds depend largely on the sizes, shapes, concentration; and phase of the cloud hydrometers (cf. Stephens et al., 1990). Second, the glaciation of lower-level clouds plays a major role in precipitation formation and hence modulates both the global hydrologic cycle and the duration and extent of global cloud cover (e.g., Wallace and Hobbs, 1977).

The role of aerosol particles in the formation of the ice phase in clouds is complex, much more so than that of CCN in forming cloud drops. This complexity arises partly because of the multiplicity of methods by which appropriate ice-forming aerosol particles (called ice nuclei) can produce ice, and partly because the formation of ice in clouds can proceed not only from primary nucleation but also from secondary processes associated with interactions of ice particles with preexisting liquid hydrometers (e.g., Mossop, 1978). In certain cloud types (marine cumulus), such secondary production dominates ice formation. Nevertheless, in a wide variety of cloud types the development of ice can be attributed to ice nuclei (Vali, 1985).

The methods by which ice nuclei can nucleate the ice phase serve as a means of dividing ice nuclei into subgroups (usually referred to as modes of ice nucleation): sublimation nuclei, immersion-freezing nuclei, contact nuclei, and condensation-freezing nuclei. Each of these modes of action requires different physical and chemical aerosol properties and hence different types of aerosol particles. This variety in mode of action is no doubt a major source of the inconsistency in ice nuclei concentration observed with different techniques (Vali, 1975). Furthermore, ice nuclei (similar to CCN) activity for each nucleation method can be expressed as a function of supersaturation or supercooling. The functional relationships can be approximated with power laws and measured exponents ranging from 4 to 12, thus suggesting very sharp decreases in activity at small supercoolings. The implied extreme rarity of ice nuclei at warm temperatures (about one particle in 10^7 at -5°C) is an additional hurdle to observations.

Indeed, ice nucleus measurements are far from reliable, and any discussions of geographic distributions of ice nuclei are quite tentative. Altitude dependence is not known, and few data from remote marine environments exist, although there are suggestions of an oceanic ice nucleus source in several data sets (e.g., Bigg, 1973; Schnell and Vali, 1976). For terrestrial locales, the situation is somewhat better in that there is considerable evidence that mineral particles act as ice nuclei (e.g., Kumai, 1951; Hobbs et al., 1971; Parungo et al., 1979). Biogenic sources of ice nuclei have also been suggested (cf. Schnell and Vali, 1972; Arny et al., 1976) and could be of importance for biosphere-atmosphere feedbacks. The global significance of anthropogenic sources of ice nuclei is unclear, but specific industrial sources have been documented (Braham and Spyers-Duran, 1974).

Another important aspect of the ice nucleus activity issue has recently arisen from consideration of the glaciation temperatures of high cirrus. Several investigators have presented arguments that the ice phase in cirrus commonly arises from the homogeneous freezing (i.e., no preexisting solid phase is present) of sulfate haze droplets (e.g., Sassen and Dodd, 1988; Heymsfield and Sabin, 1989). Indeed, Sassen (1992) has presented evidence for liquid-phase cirrus formation from volcanic aerosols. Knollenberg et al. (1993) have presented evidence that glaciation at the tops of tropical cumulonimbus clouds may also arise from homogeneous freezing of sulfate haze particles. If so, an interesting linkage between ice formation and CCN arises. The freezing temperatures of the haze droplets will depend on the solute concentrations, which in turn will be strongly modulated by the CCN and size composition. This linkage could be of considerable importance. Theoretical studies (cf. DeMott et al., 1994) suggest that cirrus formation through homogeneous freezing could produce a significantly different cloud microphysics than heterogeneous nucleation with ice nuclei. The degree of homogeneous nucleation will depend on the haze particle composition and thus on CCN, as well as the vertical distribution of ice nuclei. The climate implications of this impact of differing ice nucleation modes and hence different aerosol types on cirrus microstructure are obvious.

Recommended Process Research on Aerosols and Ice Formation in Clouds

Current understanding of the relationship between aerosols and ice formation in clouds is sufficiently incomplete to preclude the formulation of a definitive research program. Nevertheless, it is clear that the issue is sufficiently important to warrant considerable effort. The following recommendations are therefore made with the object of providing a secure foundation on which future, more comprehensive programs can be built.

1. It has not yet been well established that any of the ice nuclei measurement techniques currently employed have quantitative predictive value for ice formation in clouds. Field programs should be undertaken in which current or prospective ice nucleus measurement techniques are tested against field observations to determine if they provide correct predictions of initial ice concentrations in clouds. Implicit in this is an investigation of which mode or modes of action of ice nuclei dominate ice initiation (quite possibly, no single mode dominates under all conditions). This, in turn, will require concurrent laboratory studies to establish that the measurement techniques employed in the field measure the mode of ice initiation they are thought to measure and that this mode is relevant to atmospheric conditions. Clearly, the laboratory and field studies must be conducted concurrently and interactively.

2. Related to the above recommendation, laboratory studies of the ice nucleating activity of various well-characterized, anthropogenic aerosols should be undertaken. The same technique used to determine ice nucleating activity should then be applied to field measurements to characterize the source strengths of anthropogenic sources of a similar nature to the test aerosols. The technique(s) to be utilized for ice nucleus measurement should be based on results from the program recommended in item 1 (i.e., relevant to ice initiation in real clouds).

3. The vertical distribution of ice nuclei should be measured at various locales and under various meteorological conditions. The impact of cloud scavenging on the vertical ice nuclei distribution should be investigated for various cloud types.

4. A particular type of particle that can act as an ice nucleus through several modes is mineral dust. Indeed, dust storms have been suggested to be prolific sources of ice nuclei, and the impact of such storms (e.g.,

Saharan dust storms) can be on regional to hemispheric scales. Hence, studies of the geographic distribution of dust from these storms and their associated ice nucleating activity would be a very useful step in elucidating the geographic distribution of ice nuclei.

5. Sufficient data on homogeneous nucleation of ice in cirrus clouds should be acquired to permit some assessment of its frequency to be made and whether it is truly homogeneous (i.e., the nucleation process is not being initiated by insoluble inclusions). Further, laboratory and theoretical studies of homogeneous nucleation of ice, particularly in concentrated haze particles of the same composition as those in the upper troposphere, should be undertaken. The long-utilized standard theory has been shown to be incorrect (Pruppacher, 1995), but it is not yet clear that a satisfactory substitute is in hand.

Aerosol Process Models

Currently, the aerosol component of global climate models involves a prespecified aerosol size distribution that generally is allowed to vary only with relative humidity. Aerosol size is an important determinant of optical properties, cloud nucleating properties, and wet and dry removal rates, and a more fundamentally based treatment of aerosol size is a long-term goal in global climate models. The processes determining particle size distribution include (1) direct injection of primary particles, (2) nucleation of new particles, (3) condensation, (4) coagulation, (5) hygroscopic growth, (6) heterogeneous production in clouds, and (7) mixing of different air masses. The relative role of these processes will vary with location and other factors (season, time of day, etc.). At present, it is not even clear how the level of description of the dynamics of aerosol size distribution in a global climate model is related to the required accuracy of a radiative forcing calculation. Thus, the first need in developing aerosol process models for eventual inclusion in global climate models is to determine the sensitivity of forcing to details of the size distribution. Few direct forcing calculations of this type exist (see, for example, Pilinis et al., 1995); for indirect forcing there are no models that rely on aerosol size distribution. The parameterization developed by Ghan et al. (1993) and Chuang and Penner (1995) is used in the calculation of indirect forcing by Chuang et al. (1994) and does rely on an aerosol size distribution.

Tropospheric aerosol mass size distributions typically are observed to be dominated by two modes, with a minimum separating the two modes at around 1- to 2-μm diameter (Pandis et al., 1995). Particles in the submicrometer mode ("fine particles") have different sources, much lower rates of removal by dry deposition, much higher mass scattering and absorption efficiencies, and much higher number concentrations than the "coarse" particles of the supermicrometer mode. Consequently, the first step toward including the effects of aerosol size distribution in global models should be to capture the essential differences between these two modes.

Representation of aerosol size distribution is also important for quantifying the indirect effect of anthropogenic aerosols on cloud droplet number and reflectivity, but in this case the key separation between particle sizes occurs at a smaller diameter (by a factor of 10). If aerosol mass is mainly added to particles larger than about 0.1 µm, the number of particles that activate cloud droplets at supersaturations typical of stratocumulus clouds may not increase substantially. On the other hand, if smaller aerosol particles are formed and then grow by coagulation and condensation to sizes of about 0.1 µm, CCN concentrations will increase, with an expected corresponding increase in cloud droplet number concentrations. These processes must be represented realistically in aerosol models to predict indirect forcing by anthropogenic aerosols. In short, direct aerosol forcing requires accurate specification of particle *mass* distribution, whereas indirect aerosol forcing requires accurate knowledge of particle *number* distribution.

Removal rates are as influential as formation rates in governing aerosol concentrations. Although wet removal in rain is expected to be the principal sink for submicrometer particles, dry deposition can be an important loss process for those substances whose mass is largely in supermicrometer particles. The episodic nature of wet removal processes makes these particularly difficult to represent in models. The development of methods that could characterize the time since the last rainfall event in an air mass would aid tremendously in interpretation of aerosol observations. Obviously, an accurate representation of the amount and nature of precipitation from general circulation models is needed to represent this process in aerosol models driven accurately by such GCMs. Beyond that, however, it is important to represent the efficiency of scavenging of particles as a function of size or chemical composition. For example, more hygroscopic aerosols would be expected to be more easily scavenged than less hygroscopic ones. A process representation of aerosol removal processes will allow one to address questions such as the following: How important is it to specifically represent the less efficient scavenging of less hygroscopic aerosols in global models? If populations of different aerosols with different scavenging characteristics are necessary, how does one represent the processes leading to internal mixtures of these different aerosol types? These processes and differences in aerosols are not presently represented in aerosol models but may be needed to properly represent the response of the aerosol system to decreased or increased emission of anthropogenic aerosols.

Representation of aerosol mass distributions in current climate models has shown that direct radiative forcing is not overly sensitive (i.e., order of 20 percent) to the details of size distribution as long as the aerosol is in the efficient scattering regime (Boucher and Anderson, 1995; Pilinis et al., 1995). No current model is capable of predicting aerosol number or CCN distribution, aside from using empirical curve fits; therefore, indirect forcing cannot be handled in a fundamental manner in any model.

At present, the only available atmospheric aerosol models that explicitly include both size and chemical composition resolution have been developed in the context of urban and regional air pollution (Pilinis and Seinfeld, 1988; Wexler and Seinfeld, 1991; Pandis et al., 1993; Wexler et al., 1994). Aerosol process model development and evaluation against ambient data are, in any event, needed to test the understanding of microscale chemistry and physics relevant to radiative forcing at the global scale.

Recommended Process Research on Aerosol Models

Whereas progress has been made on three-dimensional modeling of the dynamics of urban and regional aerosols, the representation of aerosol processes in global climate models is currently nonexistent. Models for aerosol mass distributions, needed for assessment of direct climatic effects, can be developed first, drawing on the experience gained at the urban and regional scale. Models for aerosol number distribution, needed to assess indirect effects, are considerably more difficult to test, and such models do not yet exist even for the urban/regional scale.

We recommend the following:

1. The level of resolution of aerosol size and composition distributions required in global climate models of the direct effect to achieve given levels of accuracy should be assessed through comprehensive sensitivity analysis.

2. Fundamental aerosol process models aimed at aerosol number distributions should be developed for eventual use in modeling indirect effects. Such models can be evaluated initially at the urban/regional scale where sufficient data exist.

FIELD STUDIES

The ambient atmosphere is of course much more complex than the representations we can include in practical models. Concentrations vary throughout regions that the models assume to be homogeneous. Layering of air with different histories is common. Coagulation and cloud processing produce a wide variety of mixing states, so that some chemicals may exist largely as external mixtures, with a fraction internally mixed with other species. Temperature differences within a real air mass may vary much more widely than in the corresponding computational box, leading to more variable reaction rates and relative humidities. How well do models really represent the chemical and optical properties of the ambient atmosphere under a variety of conditions?

The only way to answer this question is to test models ranging in complexity from fine-scale process models to global chemical models against observations. Field measurements can be aimed at determining the adequacy with which process models embody (1) mechanisms and rates of production of aerosols from gases (relevant to both direct and indirect forcing); (2) processes controlling the evolution of aerosols, including growth, activation to cloud drops, and wet and dry removal; (3) relation(s) between aerosol optical depths and aerosol properties; (4) role(s) of specific chemical classes of aerosols, such as organics, in direct and indirect forcing; and (5) cloud-activating properties of different classes of ambient aerosols.

Closure Experiments

The ultimate goal of many experiments is to determine the precision with which models can predict certain properties of the atmosphere, This quantitative comparison requires a special kind of process study called a closure experiment. In such an experiment, an overdetermined set of observations is obtained, where the measured value of a dependent variable is compared with the value that is calculated from measured values of the independent variables, by using an appropriate process model. The model need not be a complex numerical model. It could be merely conceptual or a single theoretical relationship between two variables. The important point is that an appropriate closed set of measurement variables be selected a priori to permit a model assessment.

The outcome of a closure experiment provides a direct evaluation of the combined uncertainty of the model and measurements. Close agreement between measured and calculated results demonstrates that the model may be a suitable representation of the observed system and is appropriate for further study and testing prior to use as a component of climate forcing calculations. Conversely, poor agreement is a valuable indicator of deficiencies in the model or measurements that must be corrected before proceeding further. It is necessary to test rigorously the reliability of model output because, for example, policy implications will differ dramatically if the uncertainty in a 1.0 W m^{-2} forcing prediction is 10 percent or a factor of 2.

Closure experiments can be defined and conducted over several dimensions in space and time. Zero-dimensional (point) measurements of aerosol number concentration and chemical composition (both as a function of particle size) can be used to calculate simultaneously measured dependent variables, such as aerosol light scattering and absorption coefficients and number concentration of CCN as a function of peak supersaturation (CCN supersaturation spectrum). One-dimensional (vertical column) measurements of the vertical profile of aerosol light scattering and absorption coefficients, plus radiative fluxes, can be compared with measurements of aerosol optical thickness of the entire column and aerosol optical properties and with radiative fluxes at the top of the atmosphere. Similarly, vertical profiles through clouds of radiative fluxes and cloud droplet size distributions can be compared with measured CCN spectra below the cloud and radiative fluxes above the cloud.

Another type of one-dimensional closure study is a so-called Lagrangian process study, in which the evolution of an aerosol moving with an air mass tagged with inert chemical tracers or appropriate balloons is studied. In such studies, independent variables include initial conditions, boundary conditions (e.g., the source strength of additional material introduced into the parcel), and reaction rates, and the dependent variables are the time-dependent chemical and microphysical properties of the aerosol particles.

In vertical column closure experiments for aerosol optical depth, radiative forcing by aerosol is calculated (with suitable assumptions) for comparison with directly measured forcings. The intention of the clear-sky radiation closure experiment is to compare satellite radiation measurements and surface-based column-integrated radiation measurements with in situ (aircraft) aerosol chemical, physical, and optical measurements. An airborne aerosol lidar can be used to scale the in situ observations over appropriate altitude intervals and to identify aerosol layers that might have been missed by in situ measurements. Critical aerosol measurements in the column closure experiment include size distributions, size-resolved chemical composition, light scattering (total and hemispheric backscatter), light absorption, spectral optical depth, and vertical distributions of aerosol backscattering. Solar and infrared, upwelling and downwelling radiation, as well as meteorological/state parameters, are also needed as a function of altitude; flight profiles can be designed to observe changes in radiation levels between altitudes and relate them to size, concentration, composition, and optical thickness spectra of particles in each altitude interval. Note that some instruments determine optical depths by looking upward at the Sun, whereas other approaches permit calculation of optical depths by looking downward at the surface. It is possible, then, to seek closure among all these methods, as well as between them and in situ observations of aerosol optical properties (Russell et al., 1994).

A number of issues arise when considering optimum protocol for a column closure experiment. One fundamental issue is an appropriate horizontal scale for measurements. The importance of scale arises from disparate time constants for the various processes. Consider, for example, the linkages in Figure 2.2, all of which may be subsumed into a column closure experiment. The extreme example of scale is the closure between dimethyl sulfide (DMS) emissions and aerosol optical depth. Because DMS mixing and oxidation time scales are of the order of days, a horizontal scale of the order of 1000 km would be needed to encompass the radiative effects of those emissions. The requirement on a spatial scale is much less severe if the closure experiment were to aim only to relate existing aerosol column concentrations to aerosol optical depth.

A related scale issue in the column closure experiment concerns the altitude range of measurements. In most areas the boundary layer will contain the highest concentrations of aerosols and should therefore be sampled most intensively. The potential for layering within the boundary layer, however, and particularly within a few tens of meters of the surface in marine areas, suggests that vertical profiles of concentrations and thermodynamic and dynamic variables in marine regions need to be measured carefully. It is necessary to characterize relative humidity as a function of altitude very accurately. A maximum in relative humidity can occur at the top of the boundary layer, with concomitant effect on aerosol size. Higher concentrations of sea salt and proximity to surface sources and sinks can affect the ability of surface and shipboard samplers to represent even the lower hundred meters of the atmosphere. Because most aircraft cannot safely sample in this region, balloon-borne instruments may be required.

Similarly, the stratosphere and subsiding regions of the upper troposphere can contain significant layers of particles, thereby thwarting attempts to obtain closure between integrated in situ observations and optical depth. Mineral aerosol (dust) plumes are often visible in satellite imagery, which suggests that they have a significant radiative impact. However, their transitory nature, layered structure, and preponderance of supermicron particles make them difficult to sample from aircraft. The local nature of in situ measurements requires that serious consideration be given to the issue of representative sampling in many of the likely study regions.

Another major issue in designing closure experiments is the climatology and source distribution of proposed study areas. Process and closure experiments need to be conducted in air masses exhibiting the influence of anthropogenic emissions and in those without such influence, in both marine and continental regions. If there are gradients in sources within the study area, the experiment's design must show whether these can be exploited (e.g., by looking for gradients in optical depth with distance from a coastline) or whether they may introduce so much spatial variability that it becomes impossible for in situ measurements to represent average conditions within parts of the satellite scene.

As with direct forcing, indirect forcing is amenable to column closure experiments. In this instance, the closure variable is cloud albedo. In situ measurements of CCN, CDNC, and sulfate mass would permit the assessment of several of the linkages in Figure 2.2. These linkages, once quantified, could then be combined to derive a cloud albedo for comparison with remotely retrieved cloud albedo.

Multiplatform Field Campaigns

In many cases, a single observing platform is inadequate to test a model or determine the relative rates of competing processes. Simultaneous observations at the surface, within clouds, and from above may be required to test models of the radiative impact of clouds, for instance. This implies a need for at least three separate measuring systems. Lagrangian observations within an air mass may require measurements both at the surface and from a relay of aircraft. In each of these situations, the presence of observing systems at multiple locations allows conclusions that no single platform could provide

To evaluate our understanding of aerosol forcing of climate, both direct and indirect, multiplatform field programs will be required in which a comprehensive suite of species is measured simultaneously.[1] Various observational platforms have unique capabilities: Vertical profiles from aircraft or balloons can quantify column budgets, as well as the vertical gradients from which entrainment and deposition fluxes can be computed. Surface stations and ships are best suited to generating continuous time series of species concentrations and intensive properties, to identify the dependence on changing solar intensity levels and other environmental variables. Satellite platforms are able to measure aerosol properties over larger spatial scales than surface and airborne platforms. Satellites can therefore help define the region and global context of in situ measurements. Multiplatform experiments make it possible to employ observational strategies that supply vastly more information than each of the platforms operating independently. It is essential that such large field studies be designed for intercomparisons, so that closure between calculated and observed variables can be tested.

Surface observations of time-varying properties contain information about formation, processing, and removal rates, but are usually confounded by the effects of entrainment of air with different properties from higher altitudes or advection of new air masses to the measurement site. In situ measurements of vertical profiles from balloons or aircraft are required to factor out dynamic effects and derive processing rates.

Likewise, airborne measurements suffer from short durations of observations. Simultaneous time-series data from ships, surface sites, or coordi-

---

[1]In some cases, comprehensive field studies are not the most efficient: well-conceived studies by individuals or small groups of investigators can make significant progress on many issues. Although many questions can be addressed only with comprehensive field programs, the creativity, simplicity, and low cost inherent in small programs should be promoted.

nated aircraft can supply essential temporal information. Multiple aircraft can measure spatial gradients and follow the evolution of a study air mass in a Lagrangian reference frame. Although Lagrangian observations do not eliminate dynamic effects from entrainment or dispersion, they vastly improve the chances of factoring out these effects from the aerosol processing itself. In studies of aerosol processing in marine regions, it is frequently necessary to use simultaneous aircraft and ship observations. Whereas the aircraft can assess vertical profiles and rates of entrainment of free tropospheric air, the ship is able to make time-series measurements as well as study seawater concentrations and factors driving the exchange.

Both Lagrangian (moving air parcel) and Eulerian (fixed monitoring station) approaches can be used to address questions of aerosol formation, transformation, and removal within the framework of the large field study. Lagrangian experiments offer the potential to study oxidation processes and chemical budgets in an evolving air mass. Quasi-Lagrangian observations during ASTEX/MAGE (Atlantic Stratocumulus Transition Experiment/Marine Aerosol and Gas Exchange), NARE (North Atlantic Regional Experiment), and PEM–West (Pacific Exploratory Mission–West) suggest that this approach can yield process information while reducing confounding effects of air mass changes. An Eulerian process study can be useful at a single location, if the source field is horizontally homogeneous over sufficiently large distances that diurnal changes at one site can be assumed to represent the region. Some remote island sites may satisfy this criterion. Eulerian measurements can also be used with multiple stations, because the utility of Eulerian data depends on a trade-off between spatial homogeneity and station density.

Regardless of locations selected for large-scale studies on radiative forcing by aerosols, questions of global representativeness must be addressed. Field studies should be undertaken in regions where the fundamental controlling factors are significantly different from those of other regions. Global surveys, satellite data, and long-term monitoring can then be used to extrapolate from these representative studies to the global scale.

Two examples of multiplatform field campaigns would be those addressing the linkages between sources of anthropogenic SO_2 and sulfate aerosol, and between organic aerosols and soot from biomass burning and radiative forcing. The oxidation rates and conversion efficiencies of SO_2 by homogeneous and heterogeneous mechanisms are critical input parameters for calculating sulfate aerosol column burdens in aerosol-climate models. How much SO_2 is oxidized in the gas phase by OH, relative to oxidation in-cloud and on the surfaces of aerosols? What fraction of SO_2 is removed through deposition to surfaces before it can be oxidized to submicron sulfate aerosol? How does the presence of ammonia vapor, soot, trace metals, and condensable organics affect aerosol nucleation and growth?

One approach to answering these questions is repeated observations by multiple platforms of an air mass as it moves, for example, off the northeast coast of the United States over the North Atlantic Ocean.[2] A variety of measurements is needed to clarify the oxidation and removal pathways of this anthropogenic material. The necessary measurements include

- SO_2 and H_2SO_4, soot, organics, and trace metal concentrations;
- photochemically active trace species concentrations;
- short time-scale measurements of both sub- and supermicrometer nss (non-sea salt)-sulfate and organics;
- mass size distributions of aerosol chemical species;
- number size distributions from 3-nm to 10-µm diameter; and
- dynamical factors such as entrainment rates, turbulent transport to and from the surface, and mixing depths.

The partitioning of SO_2 oxidation products and organics between new particle production and particle growth affects the submicron aerosol size distribution and, in turn, the impact of these particles on forcing. Parameters that determine whether gaseous species nucleate to form new particles or condense onto existing particles include the saturation vapor pressure of the condensing species (H_2SO_4, NH_3, organics), relative humidity, temperature, and the existing particle number size distribution. Measurements of the size distribution between 3- and 10-nm diameter should be used to identify events of new particle production. By combining these measurements with observations of gas-phase aerosol precursors, it is possible to directly compute aerosol nucleation and growth rates. Chemical mass size distributions, the ammonium to nss-sulfate molar ratio as a function of size, and single particle analyses can help determine the role of specific chemical species in new particle production. These same measurements can define the conditions that inhibit new particle production.

The first major, multiplatform field campaign aimed at aerosol forcing of climate was ACE-1 (IGAC, 1995a). The Southern Hemisphere Marine Aerosol Characterization Experiment (ACE-1) attempted to quantify the combined chemical and physical processes controlling the evolution and properties of the atmospheric aerosol relevant to radiative forcing and climate. The goal of ACE-1 was to document the chemical, physical, and optical characteristics and determine the controlling processes of the aerosol in the remote marine atmosphere. ACE-1 was conducted from November 15 to December 14, 1995, over the southwest Pacific Ocean,

---

[2]Lagrangian measurements of polluted air passing off a continent are conceptually appealing to evaluate aerosol removal processes but are extremely difficult in practice (Slinn et al., 1983). Once the major source terms have been exhausted, the problem is simplified considerably; the reduction in aerosol concentrations may then be dominated by removal (deposition) and dilution processes, such as mixing with cleaner air. Since continental-outflow plumes are often both layered and highly heterogeneous spatially, Lagrangian experiments need to include sufficient meteorological data to trace wind trajectories at a number of vertical levels in the atmosphere.

south of Australia, and involved the joint efforts of the International Global Atmospheric Chemistry (IGAC) Project's Multiphase Atmospheric Chemistry (MAC) Activity and Marine Aerosol and Gas Exchange (MAGE) Activity. The Tropospheric Aerosol Radiative Forcing Observation Experiment (TARFOX) (June 1996) will focus specifically on the column-integrated direct radiative forcing by anthropogenic aerosols on the east coast of the United States. ACE-2, the North Atlantic Regional Aerosol Characterization Experiment, scheduled for 1997, is the third experiment coordinated by IGAC that addresses the properties of the atmospheric aerosol relevant to radiative forcing and climate. ACE-2 will extend these characterization and process studies to the North Atlantic Ocean with an emphasis on the anthropogenic perturbation of the background aerosol. A major focus of ACE-2 will be the characterization and evolution of anthropogenic aerosols from the European continent and desert dust from the African continent, as they move out over the North Atlantic Ocean.

Recommended Field Studies

This panel expects that a multiplatform field campaign of the scope of ACE-1 and ACE-2 will be needed approximately every two years over the next decade. These campaigns will be directed at understanding processes and testing models in the clean marine atmosphere, over continents, in a polluted marine region, and in the presence of biomass burning products. It is important that detailed observations be made in each region where differences in either aerosol composition or atmospheric dynamics are likely to cause models to misrepresent the aerosol radiative forcing.

SATELLITE OBSERVATIONS AND CONTINUOUS IN SITU MONITORING

Aerosols have been monitored for decades for a multitude of purposes relating to their effects on health, visibility, acidification, corrosion, and climate. Unfortunately, each effect relates to different properties of the aerosol, and the monitoring strategies used to study various effects are often incompatible. Aerosol monitoring programs to date have focused almost exclusively on extensive aerosol properties, and systematic observations of intensive aerosol properties are sorely lacking. Consequently, despite the enormous effort that has gone into studying aerosols, much remains to be learned about the spatial distributions, seasonal variability, and long-term trends of the radiative, microphysical, and chemical properties of atmospheric aerosols that determine their effects on climate. For example, the values of aerosol single scattering albedo and asymmetry factor (see discussion of optical properties above) used in radiative transfer models, both for climate forcing calculations and for satellite data retrieval algorithms, are very poorly constrained by observations.

No single approach to observing atmospheric aerosols will provide the data necessary for monitoring all important variables at all relevant spatial/temporal scales. In situ (ground-based, shipborne, and airborne) observations can provide detailed aerosol characterizations but only for limited spatial scales. Remote sensing (from satellites, aircraft, and the surface) can provide a limited set of aerosol properties up to global spatial scales but cannot provide the chemical information needed for closure with global chemical models. Fixed ground stations are suitable for continuous observations over extended time periods but lack vertical resolution, although lidars can provide useful, continuous information on the vertical distribution of aerosols. Aircraft and balloons can provide comprehensive aerosol characterizations through the vertical column, but not continuously. Only when systematically combined can these various types of observations produce a data set from which point measurements can be extrapolated with models to large geographical scales, satellite measurements can be compared with results from large-scale models, and process studies can permit general conclusions from experiments conducted under specific conditions.

Evaluation of model predictions and remote-sensing algorithms is based on spatial and temporal distributions of aerosol properties; these properties are also used directly to evaluate trends, effects, and responses to changes in emissions. With explicit recognition of the different uses of extensive and intensive aerosol properties, specific objectives of the aerosol monitoring component are (1) to determine spatial distributions of relevant extensive aerosol properties on a global scale, along with their temporal trends and seasonal cycles; and (2) to determine means, variabilities, and trends of relevant intensive aerosol properties for key aerosol types.

The monitoring data can be used in two ways to evaluate the model predictions. In the first method, the observed concentrations and deposition from in situ measurements (means and variances) and/or satellite-derived aerosol optical depth are compared with results from a model whose wind and precipitation fields are derived from a climate model. In this case, data from several years are desirable, because even monthly averaged concentrations can vary by a factor of 2 from one year to the next (e.g., Galloway et al., 1992). The value of this method lies in the ability to evaluate all aspects of a model that will be ultimately used to derive the climate response to aerosol forcing. In the second method, the observations for specific days are compared to results from a model whose winds and precipitation fields are derived from either analyzed fields or predicted fields from a weather prediction model (e.g., Benkovitz et al., 1994). This technique is especially useful for evaluating predictions of the chemical transport, transformation, and deposition characteristics of a model, although it is subject to uncertainties because precipitation is often poorly characterized by weather prediction models.

The need to integrate measurements from a variety of observational platforms with the results of process studies and numerical models dictates that standardized sampling protocols be employed. Aerosol optical properties are strongly dependent on particle size, which in turn is strongly dependent on relative humidity. For certain types of climatological observations, both relative humidity and the size range of particles sampled must be controlled if reproducible and provable comparable results are to be obtained. *Currently, no generally accepted standards exist for aerosol measurements directed toward climate forcing questions.* As a consequence, the available data from different studies often cannot be compared directly. Standard methods have been developed for use in health effects research and have been used for other atmospheric aerosol research topics as well, but the sampling criteria for health effects studies (primarily penetration into the lungs) are very different from criteria appropriate for climate studies (efficiency for light scattering, cloud nucleating properties). One of the first tasks should be to develop standard sampling protocols to be used in monitoring programs, particularly for relative humidity and particle size ranges.

Satellite Remote Sensing of Aerosols

The primary objective of satellite-based observations of aerosols is to provide a global, vertically resolved climatology of aerosol extinction throughout the troposphere and stratosphere. Also, where possible, these observations will be used to characterize additional aerosol properties such as composition, mass and surface area density, and effective radius.

Satellite-based observations of atmospheric aerosols commenced in 1978 with the Stratospheric Aerosol Measurement (SAM II) and have come to include a variety of instruments that use almost exclusively passive techniques to measure one or more properties of aerosols. A passive technique uses solar radiation that has been scattered or absorbed by aerosols, or infrared radiation emitted by aerosols, as a basis to deduce aerosol properties. SAM II, the Stratospheric Aerosol and Gas Experiment (SAGE II), and the Halogen Occultation Experiment (HALOE) comprise one set of these instruments that measure the transmission of solar radiation through the limb of the atmosphere. Since these instruments are calibrated by measuring the unattenuated Sun during each measurement event, this approach is well suited to long-term trend measurements with high vertical resolution. At the same time, spatial and temporal coverage is determined and limited by spacecraft orbital characteristics. Another limb viewing instrument, the Cryogenic Limb Array Etalon Spectrometer (CLAES), used aerosol infrared emission to infer aerosol extinction profiles. Infrared emission of aerosol is dependent on the composition and mass of aerosol but only weakly dependent on aerosol size distribution. Usually, measurements in the infrared are limited to cases of enhanced aerosol loading such as in the presence of polar stratospheric clouds (PSCs) or in the aftermath of a major volcanic eruption such as Mt. Pinatubo, and cannot provide aerosol information at lesser values of aerosol loading. Tropospheric measurements using limb techniques are complicated, but not necessarily prohibited, by the presence of horizontally inhomogeneous clouds. For instance, the SAM II/SAGE series of instruments have provided a good climatology of tropospheric aerosol extinction above 6 km, especially poleward of the tropics. In contrast, nadir-viewing passive instruments such as the Advanced Very High Resolution Radiometer (AVHRR) provide high horizontal and temporal resolution, but the measurements are limited to columnar measurements of aerosol optical depth over ocean where the surface albedo is relatively constant and well known. In addition, the retrieval process for this technique requires substantial modeling of the surface optical properties and the aerosols themselves. In September 1994, the first lidar flown in space for atmospheric studies, the Lidar In-Space Technology Experiment (LITE), demonstrated the utility of active measurements of aerosols (Figure 2.3). During a 10-day shuttle mission, high-resolution vertical and horizontal profiles of tropospheric aerosol backscatter were obtained, often in the presence of overlying high clouds.

Currently, long-term, satellite-based aerosol data sets are limited to the 17-year climatology of stratospheric and upper tropospheric aerosol properties based on SAM II (1978-1994) and on SAGE and SAGE II (1979-1981 and 1984-present, respectively). Aerosol optical depth and aerosol characteristics such as surface area and density derived from this data set have already provided significant insight into volcanic effects [including aerosol forcing (McCormick et al., 1995)] and the long-term stratospheric ozone trend (Solomon et al., 1995). Unfortunately, there is no tropospheric data set that is comparable for duration and accuracy—requirements for aerosol forcing studies.

Future spaceborne sensors must face the reality that the derivation of aerosol properties is mathematically underdetermined such that it is impossible to unambiguously derive a complete description of atmospheric aerosols from satellite-based measurements. The type and quality of aerosol information that can be derived are dependent on the technique and instrument. For example, SAGE II multiwavelength extinction data can be used to derive profiles of aerosol surface area and density and effective radius but not total particle number (Thomason and Poole, 1993). Similarly, AVHRR data can be used to derive column optical depth in the midvisible but only over ocean and for somewhat enhanced aerosol levels (Ignatov et al., 1995).

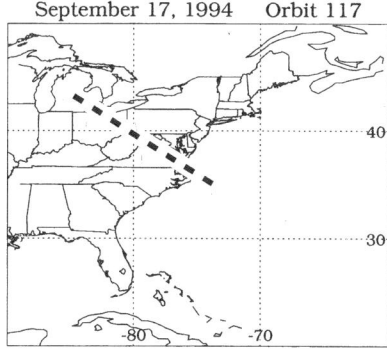

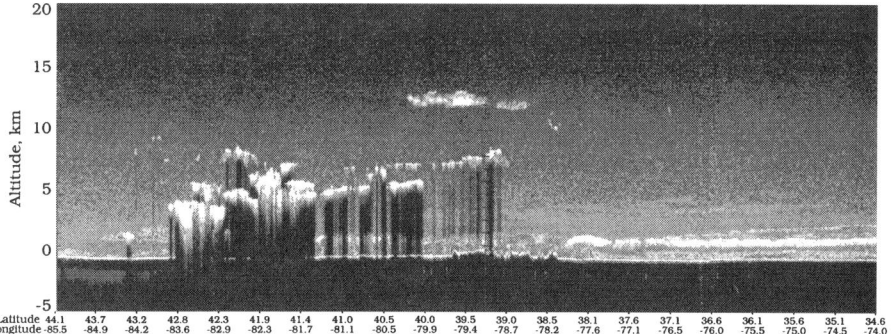

FIGURE 2.3 Observations of continental haze by LITE (Lidar In-Space Technology Experiment). The dashed line extending from central Michigan to the north of Cape Hatteras shows the path of the satellite from which the observations were made.

Accurate in situ observations of aerosol provide essential constraint to satellite retrievals and result in the best depiction of aerosol properties. In fact, it is obvious that both in situ and satellite observations by several instruments are required to adequately address the aerosol climate issue.

Table 2.2 displays currently planned and proposed satellite instruments that have the potential to yield information on tropospheric and stratospheric aerosols. These systems are clearly limited in regard to their application to understanding the climate forcing problem. The techniques employed by the nadir-viewing passive systems are expected to work only over ocean, or to experience significant degradation over land, and to provide no data on stratospheric aerosol or in the presence of cloud. In addition, retrieval schemes for passive nadir instruments almost always require extensive a priori modeling of surface properties and the aerosols themselves. Finally, inasmuch as optical depths of the order of 0.05 must be measured to an accuracy of 10 to 20 percent, only SAGE III in the stratosphere and spaceborne lidar (SPARCLE, the Spaceborne Aerosols and Cloud Lidar Earthprobe) in

TABLE 2.2 Satellite Instruments

Instrument and Mission	Launch Date	Measurement Technique	Spectral Bands	Primary Aerosol Products	Probable Resolution in Optical Depth
LITE/Shuttle (Lidar In-Space Technology Experiment)	1994	Active lidar	1064, 532, 355 nm	Backscatter profile 15-m vertical resolution	20% of value
GOME/ERS-2 (Global Ozone Monitoring Experiment)	1995	Passive, nadir viewing	240-790 nm	Optical depth	0.05
SeaWIFS (Sea-Viewing Wide Field of View Sensor)	1995	Passive, nadir viewing	412, 443, 490, 510, 555, 670, 765, 865 nm	Optical depth, over ocean	0.03
AVHRR/NOAA-K (Advanced Very High Resolution Radiometer)	1995	Passive, nadir viewing	630, 900, 1590-1780 nm	Optical depth, over ocean	0.05
POLDER/ADEOS (Polarization and Directionality of the Earth's Reflectances)	1996	Passive, nadir viewing with polarization measurement	443, 490, 565, 665, 762, 765, 865, 910 nm	Optical depth, over ocean, possibly over land	
ILAS/ADEOS (Improved Limb Atmospheric Spectrometer)	1996	Passive, solar occultation	1024 channels, 753-784 nm; 44 channels, 6-12 µm	Extinction profile, above polar regions, 2-km vertical resolution above 15 km	

Instrument	Year	Type	Wavelengths	Measurement	Accuracy
MODIS/EOS-AM (Moderate-Resolution Imaging Spectroradiometer)	1998	Passive, nadir viewing	19 channels, 400-2130 nm	Optical depth, over ocean	0.03
MISR/EOS/AM (Multiangle Imaging Spectroradiometer)	1998	Passive, nadir viewing	440, 550, 670, 860 nm	Optical depth, over ocean	0.05
SCIAMACHY/ENVISAT-I (Scanning Imaging Absorption Spectrometer for Atmospheric Cartography)	1999	Passive, nadir viewing, limb scattering, occultation	240-2380 nm	Optical depth, extinction and scattering profiles	0.05
SAGE III/EOS (Stratospheric Aerosol and Gas Experiment)	1998, 2001, TBD	Passive, solar and lunar occultation	290-1550 nm	Extinction profile, stratosphere and upper troposphere, 1-km vertical resolution	±5% at 0.001-0.1
SPARCLE (Spaceborne Aerosol and Cloud Lidar Earth Probe)	TBD	Active, lidar	532 nm with depolarization	Backscatter profile, 15-m vertical resolution	±5% at 0.1 ±50% at 0.03
EOSP/EOS-AM2 (Earth Observing Scanning Polarimeter)	2003	Passive, nadir and limb viewing with polarization measurement	12 channels, 410-2250 nm	Optical depth	0.03

NOTE: TBD = to be determined

the troposphere, given supporting in situ measurements, can meet the required accuracy demands. Nonetheless, the diversity of current and future spaceborne instruments, such as MODIS (the Moderate-Resolution Imaging Spectroradiometer) and MISR (the Multiangle Imaging Spectroradiometer), is expected to enhance overall understanding of tropospheric aerosol characteristics throughout the atmosphere. Undoubtedly, the most significant shortcoming in the current and planned suite of instruments is the lack of any instrument designed specifically to profile tropospheric aerosols. This task requires new and innovative techniques such as a spaceborne lidar system or some as yet unspecified approach.

In order to fulfill the need for a global data set of aerosol measurements, we believe that a limb instrument operating in visible and near-infrared wavelengths, such as SAGE III, in an inclined orbit will provide the stratospheric data required. However, the next flight of SAGE III in this type of orbit does not occur until 2001 or later. This virtually ensures a substantial break in the stratosphere aerosol climatology initiated by SAM II in 1978. We strongly support a flight of opportunity for SAGE III at the earliest possible date in an inclined orbit (like SAGE II's). In addition, we believe that spaceborne lidar offers the greatest likelihood of producing the data set required to understand the impact of tropospheric (and anthropogenic) aerosols on climate. The strengths of such a system relative to other nadir-viewing instruments include high vertical and horizontal resolution; an indifference to land/ocean effects; only a second-order model influence in the retrieval of the backscatter profile (estimation of the extinction-to-backscatter ratio); and the ability to measure between, and often through, clouds. Since it is unlikely that cross-track scanning will be a part of any first-generation spaceborne lidar, the cross-track horizontal resolution is reduced relative to nadir imagers such as AVHRR and MODIS. The combination of a carefully planned in situ measurement network, and a satellite platform with a cross-track scanning imager sensitive to aerosols and a spaceborne lidar, such as SPARCLE, would be the required approach for depicting global tropospheric aerosols.

In Situ Monitoring of Aerosols

Satellite measurements are appealing because of their global coverage. However, satellites alone cannot yield the necessary chemical or microphysical data. The strength of surface-based, in situ observations is in providing temporally continuous and detailed information on chemical and microphysical properties; their weakness is limited spatial coverage: in situ measurements at fixed sites can cover only a limited geographical area and only at the surface. Surface-based remote sensing techniques permit measurements of aerosol optical depths and vertical profiles of aerosol backscattering. Adding these measurements to in situ surface measurements allows ground-level results to be placed in context with vertical variations.

In situ observations provide mean values and temporal variability of key parameters required by global-scale models. Measurements of intensive aerosol properties at a number of sites, carefully stratified according to aerosol type (by using air mass trajectories and aerosol chemical composition, as well as gas-phase tracers of anthropogenic sources), can guide choices for numerical values of key parameters for models. Additionally, spatial distributions determined from a network of ground-based measurements, supplemented by regular horizontal transects obtained from aircraft and by regular vertical profiles obtained from aircraft and balloons, allow large-scale closure experiments on three-dimensional distributions of chemical species calculated from models. Such a set of observations is also needed to develop and validate the algorithms that are used to derive aerosol properties from satellite remote sensing data, as well as the algorithms that are used to remove the confounding effects of aerosols from other remotely sensed properties (e.g., sea surface temperature).

Two different strategies for in situ observations are required to address different uses of intensive and extensive aerosol properties. Measurements of a few extensive aerosol properties, with aerosol optical depth having the highest priority, are needed at a sufficient number of locations to provide the geographical variability suitable for evaluating predictions of chemical transport and radiative transport models. Additional extensive properties (such as size-resolved mass concentrations for dominant chemical species) are highly desirable, but the cost and complexity of the measurements limit the number of sites at which these could be made. Instead, efforts should be made to use data from existing networks to validate model predictions of the spatial and temporal distribution of the mass concentrations of important chemical species. In a similar vein, existing networks that monitor aerosol wet deposition could provide data for validating model predictions of removal fluxes.

Much more complex instrumentation is required for measuring intensive aerosol properties, requiring a different monitoring strategy. Table 2.3 lists aerosol parameters that we recommend be monitored continuously to determine the most important intensive aerosol properties. Table 2.4 lists the categories of sites that are needed. The recommended measurements can provide a continuous time series of all intensive aerosol properties needed for calculating aerosol radiative forcing, except angular scattering function (or asymmetry parameter) and humidity dependence of aerosol light absorption. Methods for direct determination of the latter two do not presently exist, and instruments to determine the angular scattering function are not commercially available for routine monitoring applications. Inclusion of the backward hemispheric component of aerosol light scattering will

TABLE 2.3 Aerosol Properties Needed at Continuous Monitoring Sites

Description	Recommended Approach
Mass concentration of important chemical species including major ions, organic carbon, elemental carbon, and trace elements; the total mass concentration should also be determined	Impactor/filter sampler; ion chromatography, combustion, PIXE.[a] Measurements should be obtained in two size fractions (sub- and super-micrometer)
Scattering, hemispheric backscattering, and absorption components of the aerosol light extinction coefficient, in the range 0.35-0.90 μm	Integrating nephelometer, continuous light absorption photometer
Total number concentration	Condensation particle counter
Vertical profile of aerosol backscatter	Lidar
Aerosol optical depth at ~5 wavelengths in the range 0.35-0.90 μm	Tracking sun photometer or shadowband radiometer
Surface radiation budget	Pyranometer, pyrgeometer, pyrheliometer
Hygroscopic growth factor for scattering and hemispheric backscattering components of the aerosol light extinction coefficient	Humidity-controlled integrating nephelometer
Number size distribution, 0.05-5 μm diameter	Differential mobility analyzer, optical particle counter, aerodynamic particle spectrometer

NOTE: Measurements of hygroscopic growth factor and number size distribution are not required continuously. These can be surveyed intermittently with a sampling package that circulates among the sites.

[a]PIXE = proton-induced x-ray emission analysis.

at least provide a substantial constraint on the range of possible values for the asymmetry parameter. It may become appropriate to add measurements of CCN spectra at a later date, but considerable process research and instrument development are necessary before a suitable sampling strategy can be devised.

These different strategies for continuous measurements of intensive versus extensive aerosol properties dictate different station densities. A high-density network is needed for aerosol optical depth, whereas a limited number of sites, each located in an area dominated by a different aerosol type, is appropriate for monitoring intensive aerosol properties. Model results of

TABLE 2.4 Categories of Sites to Monitor Intensive Properties

Category	Aerosol and Sampling Characteristics
Polluted continental	Industrial and other anthropogenic aerosols
Polluted marine	Anthropogenic aerosols, sampled after one or more days of transport and transformation after leaving the source regions
Clean continental	Characterize continental aerosols under conditions of minimal anthropogenic influence, for comparison with polluted conditions; given the wide variety of continental conditions (forest, desert, plains), it will be necessary to sample at a number of sites, although perhaps not simultaneously
Clean marine	Characterize marine aerosols under conditions of minimal anthropogenic influence, for comparison with polluted conditions
Biomass combustion	Aerosols formed by combustion of biomass
Mineral dust	Windblown dust, sampled after one or more days of transport and transformation after leaving the source regions
Free troposphere	Characterize aerosols above the boundary layer under conditions of minimal anthropogenic influence, for comparison with polluted conditions
Stratosphere	Surface-based monitoring impossible; primary in situ monitoring effort should be balloon-based measurements of the vertical profile of aerosol size distribution

Benkovitz et al. (1994) yielded a characteristic autocorrelation distance for sulfate column burden over the Atlantic Ocean of around 1000 km. To characterize aerosol optical depth distributions over North America, where even shorter characteristic autocorrelation distances are expected because of the proximity to anthropogenic sources, would require a network of at least 30 stations. The now-defunct aerosol optical depth network operated by the National Oceanic and Atmospheric Administration (NOAA) in the 1960s and 1970s (Flowers et al., 1969) had about 40 stations spanning the United States. A similar estimate of the number of stations needed to characterize the aerosol intensive properties cannot be made, because of the lack of data or model results on the expected variability. Instead, the recommended strategy is to begin with one station in each category and use the results from the initial stations and the airborne survey flights (see discussion of mobile platforms below) to decide whether additional stations are needed.

Although many local, state, national, and international monitoring networks include aerosol measurements, many of the available data are of limited utility for climate forcing calculations because particle sampling lacks an appropriate upper size limit. Measurements of total suspended particulate mass and measurements made according to the U.S. PM-10 standard (particulate matter of diameter less than 10 μm) are largely unsuitable because they include particles that often dominate total mass but contribute little to optical or cloud-nucleating properties. Furthermore, production mechanisms for submicrometer and supermicrometer particles are very different, and a total sample will contain contributions from both types of sources. Such uncontrolled measurements may be useful in some cases for comparing measured mass concentrations of specific chemical species with model predictions, but extreme care must be taken to ensure that all sources contributing to the total sample are modeled.

This problem is somewhat reduced, but not eliminated, with measurements made according to the U.S. PM-2.5 standard, which excludes particles larger than 2.5 μm in diameter. Extensive aerosol properties determined with this size cut are suitable for comparisons with model predictions of aerosol mass concentrations. These data are also useful for determining aerosol mass balance by chemical species, which can be used to estimate relative contributions of different species to light extinction. Measurements of mass scattering or absorption efficiencies made with a size cut larger than about 1-μm diameter, however, are very sensitive to local variations of supermicrometer particles.

Measurements of the surface radiation budget (upward and downward, solar and terrestrial radiation) are recommended at sites where the full complement of aerosol properties is measured. Aerosol changes have been indicted as major contributors to changes in the radiation budget, and dedicated networks exist for monitoring changes in solar radiation reaching the surface (WCRP, 1991). These networks, however, are not measuring aerosol properties; thus, interpretation of the effects of aerosols on results from these networks is severely limited. By complementing fully instrumented aerosol monitoring sites with radiation budget measurements or by co-locating new aerosol monitoring sites at existing radiation budget [e.g., Atmospheric Radiation Measurement (ARM) Program] sites, it will be possible to evaluate the effects of aerosol on the surface radiation budget at least at a few locations.

Measuring the surface radiation budget together with aerosol properties is needed, also, to evaluate effects of changes in aerosol on cloud optical properties. Satellite observations of cloud albedo and the radiation budget at the top of the atmosphere, plus surface-based observations of aerosols and the radiation budget, are expected to permit initial estimates of the sensitivity of cloud optical properties to below-cloud aerosol.

In addition to obtaining long-term measurements, the continuous monitoring program must also include a research component into the way monitoring is conducted. Use of standardized sampling protocols is essential for obtaining comparable results, but no single protocol is optimal for all conditions and chemical species. For example, humidity-controlled sampling is necessary to ensure that the results are not controlled by variations in atmospheric relative humidity. However, the process of lowering the relative humidity also can affect the chemical composition of the particles, possibly introducing systematic biases into the measurements. The magnitude of such effects must be evaluated over the broad range of conditions encountered at the sampling sites, and revised or supplemental sampling protocols must be developed to remove any such biases.

The measurement strategy also needs a periodic evaluation of the results of the monitoring program, to address the questions of the duration of measurements and necessary changes in network density or experimental approach. Meteorological variability from year to year suggests that an initial observational period of 5-10 years is needed, and may have to continue even longer if significant trends or large variability in the results is observed.

Recommended Surface-Based Monitoring Programs

We recommend establishing a dual-density network of surface-based stations for continuous aerosol monitoring, consisting of a high-density network of ca. 30 stations for aerosol optical depth and a low-density network of 7 stations (Table 2.4) to provide detailed information on means, variability, and trends of key aerosol radiative, chemical, and microphysical properties (Table 2.3) for different aerosol types.

Mobile Platforms

The notion of regularly using a single aircraft to support surface observations in a nearly continuous circuit of geographically distributed sites is both novel and very attractive. Such an airplane could operate essentially continually, shuttling to sites across the United States with a roughly weekly schedule. Instrumentation on the aircraft should be kept fairly simple: a suite of spectral radiometers; in situ measurements of aerosol scattering, hemispheric backscattering, and absorption coefficients; and size-resolved chemical samplers. This suite of measurements covers the most important extensive and intensive aerosol properties. By shuttling between surface monitoring sites (where cooperative vertical soundings would be made), not only would the surface program be routinely augmented with soundings, but observation over a unique geographical-scale component could be added to the program.

Similarly, a routine monitoring program from ships would add a valuable component to model validation studies (Figure 2.4). This approach is

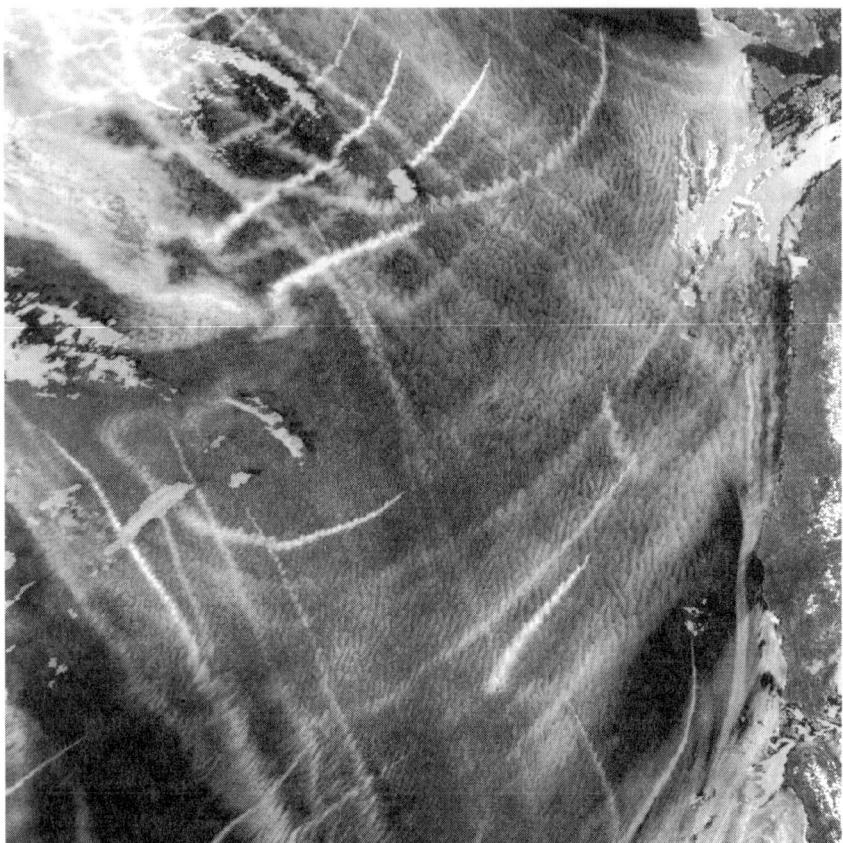

FIGURE 2.4 Ship tracks off the coast of Northern California.

an integral part of flask-sampling networks for monitoring atmospheric greenhouse gases from ships carrying freight along regular routes. For logistical reasons, the suite of measurements would have to be quite limited. As a minimum, size-resolved samples and scattering coefficients of the aerosol should be collected in fixed, predetermined regions along the track, with subsequent analyses for total mass, major ions, and light absorption. If logistical constraints allow, measurements of aerosol optical depth and aerosol light scattering and absorption coefficients should also be obtained.

The recommended approach for routine monitoring of the stratospheric aerosol is via satellite. However, satellite observations must be complemented with a regular program of balloon-borne observations to provide ground-truth information. The 20-year record of vertical profiles of aerosol size distribution at Laramie, Wyoming, should be continued to satisfy this need.

Recommended Mobile Monitoring Programs

We recommend a systematic, limited-duration (three- to five-year) program of ship- and airborne surveys to characterize the horizontal and vertical distributions, over the global oceans and North America, of the same aerosol properties studied in the low-density, surface-based network. In addition, we recommend continuing the long-term record of vertical profiles of aerosol size distribution at Laramie, Wyoming.

We recommend that a minimum of one aircraft make weekly flights to systematically make vertical profile measurements of a subset of aerosol properties.

We also recommend that a least two ships each in the Atlantic, Pacific, and Indian Oceans be outfitted with continuous and flask-monitoring systems.

RECOMMENDED TECHNOLOGY DEVELOPMENTS

The proposed systematic program of closure studies (aimed at testing the internal consistency of measurements and models) will undoubtedly reveal situations in which current measurement technology is inadequate. Furthermore, there are areas in which development efforts are needed immediately to provide instrumentation that is suitable for continuous monitoring and process/closure studies. Also, the integrated program proposed herein will require continued advances in modeling capability in both computational hardware and software.

1. *Parallel Architectures*: One of the most computer-intensive calculations with regard to modeling aerosol behavior is prediction of aerosol size evolution. At present, it is not feasible to include aerosol size resolution in global-scale models; however, massively parallel computers have the potential to greatly speed calculations and perhaps eventually allow size and composition resolution in global aerosol models. Research should be initiated on the use of parallel computers for coupled atmospheric chemistry-aerosol models. Coupled with the use of massively parallel computers, numerical methods for solving the basic equations of the models should be reassessed.

2. *In Situ Measurement of Aerosol Light Absorption*: Suitable methods exist for determining the dependence of aerosol light scattering on humidity. For aerosol light absorption, however, humidity dependence has not been measured and suitable methods are unavailable. Existing methods for determining absorption in all but the most polluted situations require particle concentration on filters to produce a measurable change in transmitted light. Unfortunately, the humidity dependence of light absorption will likely be different for deposited versus airborne particles because of the optical and physical effects of contact with a filter substrate. We recom-

mend that research be initiated on the development of in situ techniques for measuring aerosol light absorption.

3. *Angular Scattering Function and Asymmetry Factor*: A few polar nephelometers, which measure the angular scattering function $\beta(\phi,\lambda)$, are in existence (Hayasaka et al., 1992; Jones et al., 1994), but none is being used for continuous monitoring or closure studies. These instruments are needed for determining the asymmetry parameter $g(\lambda)$, for which no method of direct determination exists. Size distribution measurements and numerical inversion measurements of other aerosol optical properties can be used to estimate the asymmetry factor, but thorough closure experiments for these approaches require a direct measurement of either $\beta(\phi,\lambda)$ or $g(\lambda)$.

4. *Systematic Vertical Profiling*: Interpretations of observations from surface-based facilities require assumptions about representativeness of the observations to conditions throughout the vertical column. Systematic profiles obtained from the proposed airborne observational program will allow these assumptions to be tested, but more frequent observations at surface stations will be needed to allow evaluations of the importance of vertical inhomogeneities to measured aerosol properties. Tropospheric aerosol lidars measure vertical profiles of the volume aerosol backscatter cross section β_π (Hoff, 1988; Radke et al., 1989b; Ansmann et al., 1990). Relatively constant values of β_π in the boundary layer, with lower values aloft, would support the assumption that surface-based measurements represent optical properties of the part of the vertical column that dominates the direct aerosol radiative forcing; conversely, layers aloft with higher values of β_π would indicate that surface observations may be unrepresentative. Although methods exist for estimating aerosol extinction profiles from lidar measurements of β_π (Kovalev, 1993), the relationship is ambiguous and additional measurements (e.g., from the airborne observational program) are needed to resolve the ambiguity. However, for the purpose of identifying cases in which the surface measurements may be nonrepresentative, this ambiguity is acceptable. Analysis of tropospheric data from existing operational lidars would be a logical first step.

5. *Refractive Index*: Most methods for determining the real part of refractive indices of aerosol particles require particle collection on filters for subsequent chemical or optical analysis. For some closure studies, these methods provide insufficient time and/or size resolution for adequate inference of the refractive index. New instruments for this measurement have the potential to surpass these limitations, but they cannot yet be used routinely for process/closure studies or continuous monitoring programs.

6. *Cloud Nucleating Particles*: Several different types of CCN counters are in use today. Most have their roots in diffusion cloud chambers. In the thermal gradient cloud chamber, the aerosol is processed through a channel in which wetted parallel walls are maintained at slightly different tempera-

tures. Each wall is coated with a water-absorbing material, so that saturated conditions are maintained at each surface. Because of the temperature gradient between the two walls and the nonlinear dependence of the equilibrium vapor pressure on temperature, the saturation ratio increases from unity at the walls to a maximum in the middle of the channel. To obtain a distribution of supersaturations, a sequence of supersaturations is produced for successive measurements.

The limited time and supersaturation resolution of such measurements motivated Fukuta and Saxena (1979a,b) to develop a transverse thermal gradient CCN spectrometer capable of producing a range of supersaturations typical of those encountered in natural clouds and yielding near real-time measurements of the CCN spectrum. The range of supersaturations achievable with this instrument is about 0.1 to 2 percent (i.e., the high end of those encountered in cloud environments).

Hudson (1989) developed an instrument that both increased the time resolution beyond that achievable with the transverse gradient CCN counter and extended the range of supersaturations to values as low as 0.02 percent. The instrument does not provide an absolute measurement of critical supersaturation. Instead, it must be calibrated by using particles of known critical supersaturation (i.e., known composition and size). Moreover, the use of the final size to infer critical supersaturation makes the longitudinal gradient CCN counter sensitive to anything that might cause variations in growth rates. Experimental evidence that the condensation coefficient decreases with time following the onset of cloud formation (Hagen et al., 1989) suggests that the response of the instrument might vary with cloud age.

None of the research-grade instruments that exist for determining number concentration of CCN as a function of supersaturation $N_{ccn}(S)$ is suitable for routine operation in a monitoring network or for unattended operation in an aircraft. There is a critical need for a compact, robust instrument to measure CCN spectra in a monitoring network or in airborne mode.

7. *Lightweight Samplers*: Several new sampling platforms are being developed to reduce the cost of obtaining vertical profile information. Remotely piloted aircraft, balloons, and kites are (or will soon be) used to make measurements above the surface. Unfortunately, many of the sampling devices now in use on other platforms are unsuited to these low-power environments. It is particularly difficult to collect samples for chemical analysis since the large weight and power requirements of typical filter samplers cannot be flown by these systems. An aerosol collection device optimized for low weight and power would make it possible to sample aerosols above the boundary layer from an inexpensive kite or balloon at a much lower cost than flying a research aircraft. In view of the need for data on concentrations above the surface, the development of lightweight particle samplers needs to be pursued.

8. *Buoy-Mounted Instrumentation*: Over most of the Earth (the oceans), it is impossible to make long-term in situ observations of aerosol properties. The oceanographic and meteorological community has addressed similar sampling problems by investing in the engineering necessary to make some instrumentation compatible with long-term operation of buoys. Traditionally touchy devices, such as sonic anemometers and instruments that measure salinity and some chemical species, are now routinely operating in the middle of the oceans on buoys that are visited only annually. If one were to engineer a seawater DMS probe, an optical particle counter, or related aerosol-measuring devices to work unattended for months at a time on buoys, one could vastly improve the climatology of aerosol concentrations in remote regions. The engineering capability already exists and simply needs to be applied to the most important sensors.

SUMMARY OF A RESEARCH PROGRAM ON AEROSOL FORCING OF CLIMATE

We recommend an integrated program of research on aerosol forcing of climate that includes

1. advances in the representation of aerosols in global climate models, particularly with respect to indirect climatic effects;
2. laboratory, theoretical, and field research on aerosol optical properties;
3. identification of aerosol molecular composition, particularly the organic fraction;
4. development of an understanding of aerosol formation and growth in the atmosphere;
5. elucidation, through laboratory, theoretical, and field studies, of the aerosol-CCN-cloud droplet-albedo relationship;
6. execution of atmospheric closure experiments to test theoretical understanding;
7. development of a new satellite system for remote sensing of tropospheric aerosols;
8. establishment of in situ aerosol research measurement stations to provide continuous data on aerosol amounts and properties in key global areas;
9. advancement of instrumentation technology for measuring aerosol properties in situ; and
10. system integration and assessment.

Addressing these needs will require a systematic and patient approach. A crash program is not called for; rather, a systematic development of capabilities should be pursued over a period of the order of a decade.

3

Sensitivity/Uncertainty Analysis and the Setting of Priorities

In this chapter, the goal is to present a framework capable of focusing disparate elements into a coherent and cost-effective whole (i.e., an integrating framework for setting priorities). This framework is constructed via sensitivity/uncertainty analyses, and we call the resulting program the Interagency Climate-Aerosol Radiative Uncertainties and Sensitivities (ICARUS) Program.

The proposed ICARUS program would be an additional major component of a larger program dealing with climate change, which in the United States is known as the Global Change Research Program (USGCRP). The goal of the USGCRP is "to gain a predictive understanding of the interactive physical, geological, chemical, biological, economic, and social processes that regulate the total Earth system, and, hence, establish a scientific basis for national and international policy formulation and decisions relating to natural and human-induced changes in the global environment and their regional impacts." The ICARUS program should be incorporated and administered as a part of the USGCRP.

As a part of the USGCRP, priorities for ICARUS would be required to fit within the "prioritization scheme" used by USGCRP and would appropriately be judged relative to these existing priorities. Correspondingly, the overall budget for ICARUS would have to be consistent with the priorities USGCRP assigns to the scientific contributions required from ICARUS, weighting cost appropriately with feasibility and with the strategic and integrating priorities of the USGCRP.

Although these constraints on the ICARUS program as a part of the USGCRP would be substantial, the benefits would be enormous, including utilization of the existing USGCRP administrative structure and integration with existing USGCRP measurement and modeling capabilities.

INTEGRATION VIA SENSITIVITY/UNCERTAINTY ANALYSES

To define priorities within any research program, we accept and endorse the time-honored rational, quantitative procedure of performing sensitivity analyses and then weighting the results appropriately with feasibility and various other factors (such as cost) imposed by broader systems. However, major difficulties are encountered when sensitivity analysis is applied to climate forcing by aerosols.

To guide decisions associated with potential climate change, sufficiently reliable models of the climate must be developed. To develop reliable climate models, reliable models of climate forcing by airborne particles from future emissions are needed. Therefore, central to the ICARUS program is development of the understanding of aerosol forcing, as embodied in climate models:

- to describe, realistically, existing concentrations and properties of aerosol particles throughout the global atmosphere, thereby
- to provide climate models with needed predictions of current and future anthropogenic aerosol radiative forcings, thereby
- to meet the prime goals of the ICARUS program and the USGCRP, plus
- to establish and continuously update rational choices of ICARUS research priorities via sensitivity analyses, and
- to integrate data, theory, and applications.

AN EXAMPLE OF SENSITIVITY ANALYSIS: DIRECT RADIATIVE FORCING

We present here an example that illustrates sensitivity analysis but not uncertainty analysis (uncertainties are not evaluated) and only for direct radiative forcing by sulfate particles. The flux calculations are based on a column version of the National Center for Atmospheric Research (NCAR) CCM2 radiation model. The model atmosphere employed three layers of cloud, a high cloud layer, a midlayer cloud, and a low cloud level. The cloud amounts were adjusted to yield a top of atmosphere planetary albedo of 0.3. The lowest cloud layer is located at 800 millibars (mb). The below-cloud calculation assumed that the aerosol layer was completely below this low cloud layer. The above-cloud calculation assumed that the entire aerosol loading was located above the lowest cloud layer. Details of the radiation model are given by Kiehl and Briegleb (1993).

Table 3.1 shows the sensitivity of radiative forcing to changes in either aerosol properties or atmospheric conditions. The second column in Table 3.1 presents this sensitivity as a change in absorbed solar flux per percentage change in a property, that is, $[\Delta S_a(x + 0.2x) - \Delta S_a(x - 0.2x)] \div 0.4x$, where ΔS_a is the globally averaged forcing from an increase in sulfate aerosol and x denotes a particular aerosol or atmospheric property. Note that the control forcing ΔS_a is -0.3 W m^{-2}. The near-linear dependence of the forcing on loading is apparent from the first sensitivity (i.e., -0.3 W m^{-2} is obtained approximately by multiplying the listed value by 100 percent). The third column ranks the various sensitivities relative to the sensitivity to changes in total loading. For Table 3.1, the aerosol is assumed to reside in the boundary layer and hence below clouds. Table 3.2 shows similar results for conditions where the aerosol is placed above the lowest cloud layer (located at 800 mb). The results are shown graphically in Figures 3.1 and 3.2.

A negative sensitivity implies an increase in aerosol forcing (i.e., more negative), because the anthropogenic aerosol forcing is negative (i.e., the presence of aerosols decreases the column-absorbed shortwave flux). A positive sensitivity implies a reduction in aerosol forcing from an increase in the property. For example, an increase in the asymmetry parameter g leads to more solar radiation scattered toward the Earth and hence less back to space, which reduces the anthropogenic aerosol effect.

According to Table 3.1, for a sulfate aerosol located below clouds, the most important aerosol property affecting the forcing is the asymmetry parameter. For this parameter, the visible values are more important than the near-infrared values. Second in importance is the single scattering albedo, where the visible values are again of greatest importance. As for atmospheric properties, aerosol forcing is quite sensitive to total cloud cover.

Importantly, the ranking is dependent on the vertical location of the aerosol layer. Table 3.2 indicates that the single scattering albedo is of greater importance for a sulfate layer above the lowest cloud base. In this case, however, the forcing is still very sensitive to the asymmetry parameter. Sensitivity to other aerosol properties, such as the width of the size distribution and the chemical composition, is discussed in Kiehl and Briegleb (1993). It is important to note that the direct aerosol forcing is sensitive to total cloud cover. This places a stringent constraint on cloud predictive capabilities of global climate models and demonstrates the importance of developing ICARUS priorities within the priority framework of the entire USGCRP.

More to the present point, this analysis reveals some essential features of sensitivity/uncertainty analyses. Even though uncertainties were not

TABLE 3.1 Sensitivity Calculations for a Sulfate Aerosol Layer Below Clouds

	Sensitivity of Forcing (W m^{-2} %$^{-1}$)	Rank Relative to Loading
Aerosol Property		
Loading	-0.00297	1.00
Single scattering albedo (ω_0)	-0.00473	1.59
ω_0 (visible)	-0.00360	1.21
ω_0 (near infrared)	-0.00098	0.33
Asymmetry parameter (g)	+0.00607	-2.04
g (visible)	+0.00484	-1.63
g (near infrared)	+0.00106	-0.36
Atmospheric and Surface Properties		
Surface albedo	+0.00048	-0.16
Total cloud cover	+0.00236	-0.79
Water vapor amount	+0.00051	-0.17

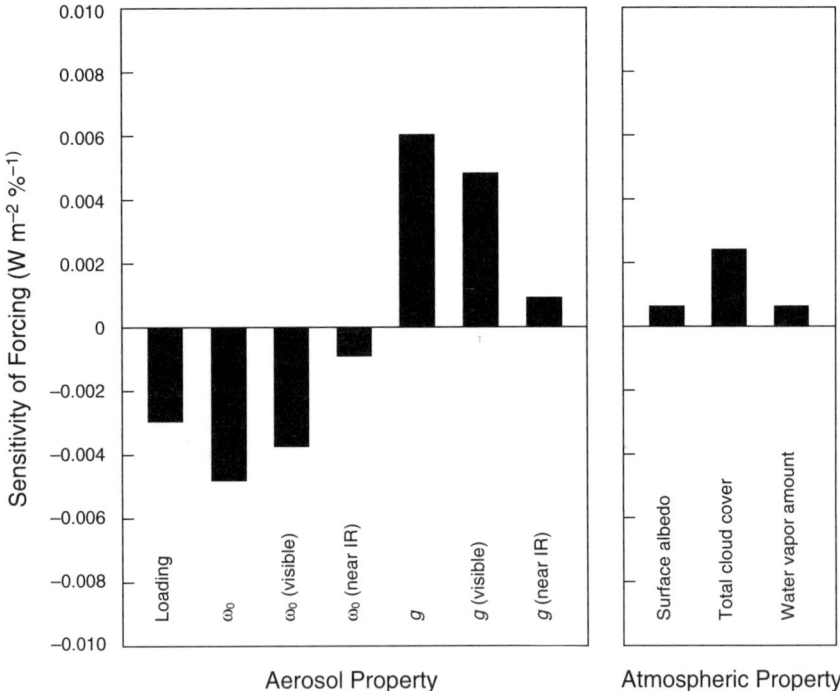

FIGURE 3.1 Sensitivity of aerosol forcing for an aerosol layer below cloud.

TABLE 3.2 Sensitivity Calculations for an Aerosol Layer Above Lowest Cloud Layer

	Sensitivity of Forcing (W m^{-2} %$^{-1}$)	Rank Relative to Loading
Aerosol Property		
Loading	-0.00227	1.00
Single scattering albedo (ω_0)	-0.00948	4.19
ω_0 (visible) ($\leq \lambda \leq$)	-0.00705	3.11
ω_0 (near infrared) ($\leq \lambda$)	-0.00202	0.89
Asymmetry parameter (g)	+0.00457	-2.02
g (visible)	+0.00358	-1.58
g (near infrared)	+0.00086	-0.38
Atmospheric and Surface Properties		
Surface albedo	+0.00035	-0.15
Total cloud cover	+0.00180	-0.79
Water vapor amount	+0.00050	-0.22

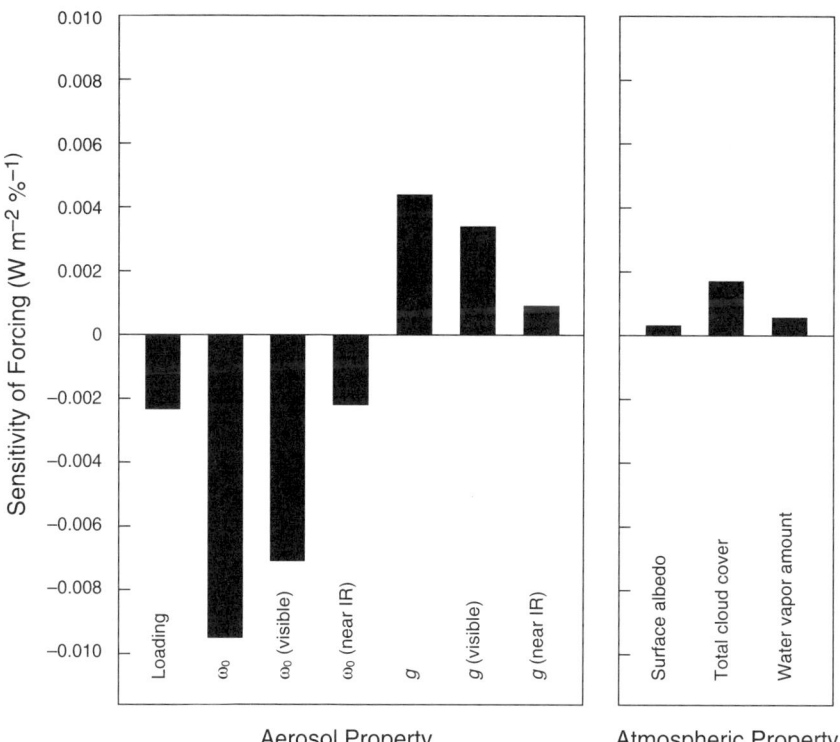

FIGURE 3.2 Sensitivity of aerosol forcing for an aerosol layer above lowest cloud layer.

explicit in this analysis, the method clearly requires quantitative descriptions of governing processes (in this case, via Mie theory of scattering). In essence, the method evaluates partial derivatives (i.e., evaluates overall changes caused by "parameter" changes).

In contrast to this illustrated case for direct effects, if quantitative, realistic descriptions of processes are unavailable (as is the case for indirect effects), then sensitivity/uncertainty analyses cannot be performed quantitatively. True, some estimates of indirect effects are becoming available, and, as reviewed in Chapter 1, emerging consensus seems to be that indirect effect are "probably comparable" to direct effects (Jones et al., 1994; Pincus and Baker, 1994; Boucher and Lohmann, 1995). From the perspective of sensitivity/uncertainty analysis, however, the reality is that the greatest uncertainty about indirect effects arises not from varying parameters within developing models but from differences among the emerging models themselves. As a result, classical sensitivity analysis is inapplicable or, at least, premature. Therefore, research priorities must be defined more heuristically, as described in the next section.

FRAMEWORKS FOR RESEARCH AND FUNDING PRIORITIES

From the previous two sections, four central features of a priority framework are apparent: (1) numerical models must be central to the ICARUS program; (2) via these models, sensitivity/uncertainty analyses can define priorities; (3) to improve the quantitative base of these analyses, an immediate ICARUS thrust must be to improve quantitative understanding of indirect effects; (4) in the meantime, until understanding of indirect effects increases substantially, priorities can be defined only on a heuristic basis. For what follows, this heuristic base is our assessment that direct and indirect effects are "probably comparable."

In this section, therefore, we present only *qualitative* analyses of research and funding priorities. We describe the results as priority *frameworks* to emphasize that complete construction currently seems impossible. To begin to outline these frameworks, consider Figure 3.3, which shows—hypothetically—relative uncertainties for the four identified aspects of the total problem: indirect effects for remote marine and continental locations, and direct effects for organic and inorganic aerosols. Also shown (again, only qualitatively) is a lower value for the uncertainties, identified as an ICARUS Phase 1 goal.[1]

---

[1] Although we have identified model development as the prime goal of ICARUS, it has not been identified separately in Figure 3.4 (and subsequent figures). The reason follows from our firm conviction that modeling must be an integral component of all aspects of ICARUS research.

SENSITIVITY/UNCERTAINTY ANALYSIS AND THE SETTING OF PRIORITIES

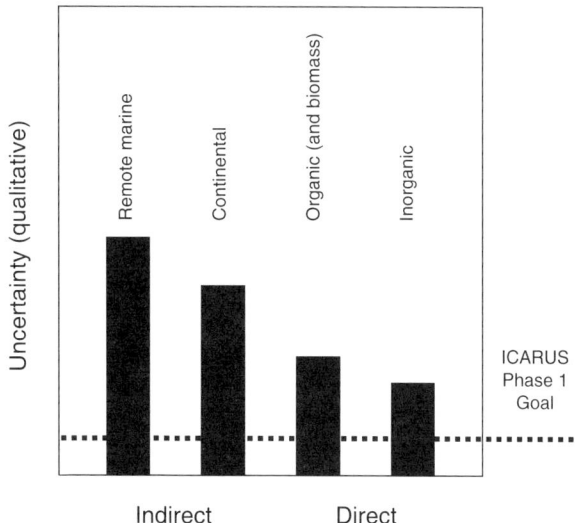

FIGURE 3.3 Qualitative indications of current radiative forcing *uncertainties* for indirect effects (separately for marine and continental clouds) and for direct effects (separately for organic and inorganic aerosols) and a qualitative indication of the uncertainty goal (to be defined by USGCRP) for the first phase of ICARUS research.

Given the uncertainty estimates shown in Figure 3.3 (presumed to be known), research priorities could be set, for example, as indicated *qualitatively* in Figure 3.4. Relative differences between corresponding bars in Figures 3.3 and 3.4 reflect, qualitatively, the weighting of the uncertainties of Figure 3.3. Typically the goal of applying these weighting factors is to progress toward solving a specific aspect of the overall problem more rapidly.

In reality, the task of "suitably weighting" other factors is nontrivial and is particularly difficult when there are conflicting views on any particular issue. Consequently, Figure 3.4 is presented only to show the method for constructing this research priority framework. Nonetheless, it might be useful to mention the following two factors that we expect would be included in such a weighting, thereby explaining the evident differences between Figures 3.3 and 3.4:

1. Because sulfur emissions especially are expected to change rapidly during the next decade (with substantially greater increases from some rapidly developing economies, such as China's, and substantially smaller releases associated with acid rain regulations in other countries), it might be supposed that, during the next decade, changes in direct radiative effects from inorganic particles (both increases and decreases) will become more

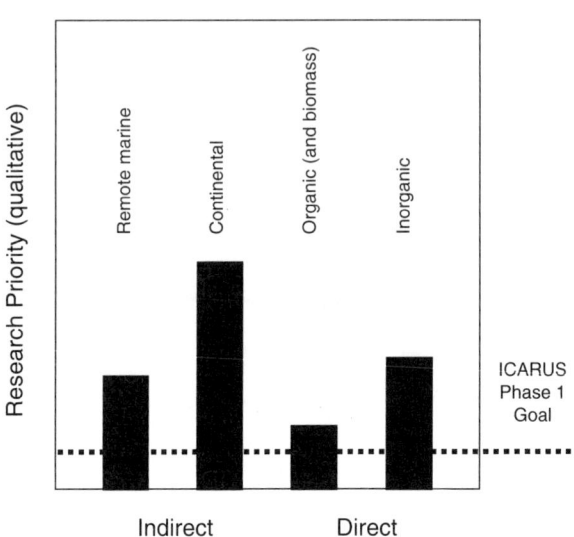

FIGURE 3.4 Qualitative indication of relative ICARUS *research priorities* for different topics, with the differences from Figure 3.3 resulting from weighting the uncertainties of Figure 3.3 with USGCRP strategic and integrating priorities; here, the weighting has been by assumed amounts.

important than those of organic particles. Consequently, for Figure 3.5, we have shown, qualitatively, an enhancement of the priority for research on the direct effects of inorganic particles as opposed to organic particles. However, we must remember that organic particles are particularly poorly understood (Penner, 1995). Hence, the research priorities shown in Figure 3.4 may indeed be reversed as more is learned about organic aerosols.

2. Because major, disruptive social actions to alleviate or anticipate global change are not likely to be undertaken until temperature data show an unambiguous warming trend; because available temperature data are primarily from continental sites; because these data currently show a decrease in the daily temperature range (DTR), caused mainly by nights warming more than days (Karl et al., 1995); because the magnitude of this DTR decrease is inconsistent with climate models that account only for increases in "greenhouse gases"; and because there are suggestions (Hansen et al., 1995) that this DTR decrease is caused by increases in low-level cloud cover (an expected indirect consequence of increasing anthropogenic aerosols)—therefore, for Figure 3.4, we have shown, qualitatively, an enhancement of the priority for research on indirect effects of anthropogenic aerosols on continental as opposed to marine clouds.

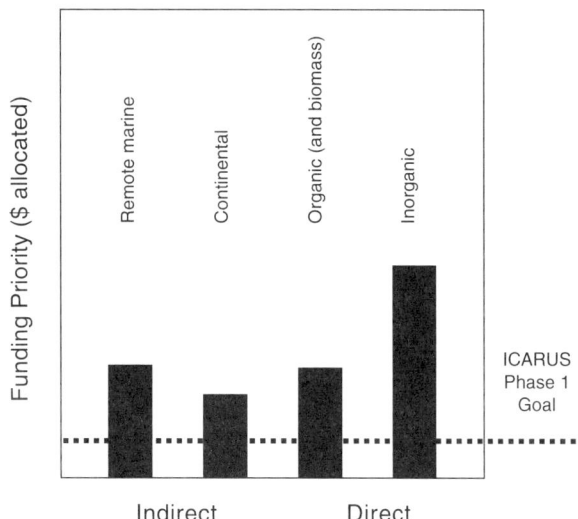

FIGURE 3.5 Qualitative indication of relative *funding priorities* (resource allocations) for the indicated broad research topics, with the differences from Figure 3.4 (research priorities) resulting from weighting these priorities with costs to perform the research; here, the weighting has been by assumed amounts.

We repeat, however, that the purpose of presenting these details (immediately above) has been only to illustrate how a research priority framework can be constructed by incorporating scientific uncertainties with other priorities, typically dictated by more comprehensive systems [USGCRP, Committee on Environment and Natural Resources (CENR), National Science and Technology Council (NSTC), et al.].

To proceed from research priorities (indicated qualitatively in Figure 3.4) to funding priorities (indicated qualitatively in Figure 3.5), it is of course necessary to estimate costs for specific research efforts within each broad research category (i.e., the four broad categories shown in Figures 3.4 and 3.5). Some of these details are indicated later in this section; first, however, to emphasize the differences between research priorities (Figure 3.4) and funding priorities (Figure 3.5), the latter figure shows a hypothetical result from such a detailed analyses.

For the hypothetical example shown in Figure 3.5, we are suggesting (only to illustrate the framework) (1) that costs to perform studies of indirect effects as they are presently done (typically performed by relatively small research groups) would normally be lower than those for studies of direct effects (and lower for continental than for marine studies, based on costs of aircraft flights and shipboard sampling); (2) that costs to study direct effects of organic aerosols would be rather substantial because of the added expense of sampling and chemical analyses (relative to well-developed techniques for inorganic aerosols); and (3) that costs to study direct effects of inorganic aerosols can be high if the decision is made to take advantage of the many benefits of remote sensing from satellites—as well as other weighting factors.

AN EXAMPLE OF SENSITIVITY AND UNCERTAINTY ANALYSES

In this section, we illustrate the sensitivity/uncertainty analysis method further by presenting both sensitivity and uncertainty analyses for direct radiative effects of sulfate particles. The goal here is to examine the framework more closely and, to this end, consider how only the rightmost bar in Figure 3.5 (i.e., costs to reduce uncertainties about direct effects of inorganic aerosols) might be derived. For this case, in contrast to the case of indirect radiative effects, considerable knowledge is already available. Also, in contrast to the case for indirect effects, there are currently about a half dozen more-or-less comprehensive numerical models that attempt to simulate these direct effects.

A partial analysis from one such model has been demonstrated above. There, results from only a sensitivity analysis were shown—for only some meteorological and aerosol property variables, and based on only a one-dimensional radiation code. A more thorough analysis would use a global atmospheric chemistry model to predict sulfate concentrations. Then, estimates could be made for the sensitivities of direct radiative forcing to uncertainties in both intensive and extensive aerosol properties and to some assumptions and approximations of the model. (If a model omits a particular process such as cloud venting, the sensitivity of the resulting radiative calculation to this omission obviously cannot be estimated.) To date (to our knowledge), such an analysis has not been performed, and because it is essential for establishing ICARUS research priorities, it will be identified as a task for immediate initiation in the ICARUS program.

In the meantime, to examine the priority framework in more detail here, we illustrate with an uncertainty analysis based on a simple, zero-dimensional model. Table 3.3 gives estimates for uncertainty factors for terms in the following equation for the areal-mean shortwave radiative forcing by sulfate and biomass burning aerosol $\langle \Delta F_R \rangle$:

$$\langle \Delta F_R \rangle = -\frac{1}{2} F_T T^2 \{1 - A_c\}[1 - R_s]^2 b \alpha_{SO_4^{2-}} f(RH) Q_{SO_2} Y_{SO_4^{2-}} \tau_{SO_4^{2-}} / A,$$

where

F_T	is the incident solar flux (solar constant times the cosine of the zenith angle (watts per square meter),
T	is the fraction of the incident light transmitted by the atmosphere above the aerosol,
A	is the area of the geographical region over which the flux is averaged (square meters),
A_C	is the fractional cloud cover,
R_S	is the albedo of the underlying surface,
b	is the upward fraction of the radiation scattered by the aerosol,
$\alpha_{SO_4^{2-}}$	is the scattering efficiency of fine-particle sulfate at a reference low relative humidity (square meters per gram),
$f(RH)$	accounts for increase in scattering with increasing relative humidity,
Q_{SO_2}	is the source strength of SO_2 (grams per second),
$Y_{SO_4^{2-}}$	is the fraction of emitted SO_2 that yields sulfate aerosol, and
$\tau_{SO_4^{2-}}$	is the sulfate lifetime in the atmosphere (seconds).

Figure 3.6 graphically shows the uncertainties estimated in Table 3.3A (for sulfate aerosols), along with a hypothetical "ICARUS Phase 1 Goal" for an acceptable uncertainty level.

At this point, it is quite understandable that disagreements would exist about the central values and ranges assigned to the quantities listed in Table 3.3 and plotted in Figure 3.6. For example, the overall uncertainty for the areal mean shortwave radiative forcing attributable to sulfate aerosol [ΔF_R] [estimated by the Intergovernmental Panel on Climate Change (IPCC, 1995a)] is 2.2, as opposed to the value 1.89 given in Table 3.3. There can be disagreements, also, about choices of entries in such tables (e.g., inclusion or omission of assumptions about atmospheric processes such as cloud venting, wet and dry deposition). More generally, there can be disagreements about the adequacies of the model(s) used to generate these uncertainty estimates.[2] Also, there is no doubt that defining an acceptable uncertainty will be difficult. Setting these difficulties aside temporarily, however, we now suggest how to proceed from Figure 3.6.

---

[2]We want to reemphasize the desirability that model diversity be maintained—and even enhanced. Given the approximations and knowledge deficiencies in so many components of existing models, this diversity is one of the few sources of strengths in existing programs. As a consequence of this diversity (and these deficiencies), resulting model predictions can differ dramatically, thereby providing a measure of appropriate confidence to be placed in predictions from any one model. Stated differently, at present there is little that would cause us greater concern about estimates of climate cooling by aerosols than community acceptance of a "standard model." Perhaps it will be reasonable to accept such a standard a decade or so from now, but we are certain that such a time has neither arrived nor is fast upon us.

TABLE 3.3 Factors Contributing to Estimates of the Direct Forcing by Anthropogenic Sulfate (A) and Biomass Burning (B) Airborne Particles, Estimated Ranges, and Resulting Uncertainty Factors (for estimates of changes in reflected solar radiation)

Quantity	Central Value	Estimated Range	Uncertainty Factor, uf_i
A. *Anthropogenic Sulfate*			
Aerosol mass scattering efficiency ($m^2 g^{-1}$)	5	3.6-7	1.400
Average atmospheric lifetime (days)	6	4-8	1.375
Aerosol hemispheric backscatter fraction	0.15	0.12-0.22	1.267
Fraction of SO_2 oxidized to sulfate aerosol	0.5	0.4-0.6	1.200
Square of atmospheric transmittance above aerosol layer	0.714	0.594-0.86	1.204
Fractional increase in scattering efficiency from hygroscopicity	1.7	1.4-2.0	1.176
Source strength of anthropogenic S (Tg per year)	71	62-81	1.141
Fraction of Earth not covered by clouds	0.39	0.35-0.44	1.128
Square of surface co-albedo	0.72	0.65-0.80	1.111

Total uncertainty factor = $\exp \{[\Sigma (\ln uf_i)^2]^{1/2} = 1.89^a$
Result: If central value is -0.6 W m^{-2}, then the range is from -0.3 to -1.1 W m^{-2}

B. *Biomass Burning Aerosols*

Emission factor (g/kg C in fuel)	32	18-65	1.750
Aerosol mass scattering efficiency ($m^2 g^{-1}$)	5	3.6-7	1.400
Average atmospheric lifetime of smoke (days)	5	3.6-7	1.400
Aerosol hemispheric backscatter fraction	0.15	0.11-0.20	1.333
Aerosol mass absorption efficiency ($m^2 g^{-1}$)	0.7	0.5-0.9	1.286
Square of atmospheric transmittance above aerosol layer	0.714	0.594-0.86	1.204
Amount of biomass burned (Tg of C per year)	3800	3200-4500	1.197
Fractional increase in scattering efficiency from hygroscopicity	1.7	1.4-2.0	1.176
Fraction of Earth not covered by clouds	0.39	0.35-0.44	1.128
Square of surface co-albedo	0.72	0.65-0.80	1.111

Total uncertainty factor = exp $\{[\Sigma (\ln uf_i)^2]^{1/2}\} = 2.45$[a]
Result: If central value is -0.8 W m^{-2}, then the range is from -0.3 to -2.0 W m^{-2}

[a]This analysis assumes that these factors are independent, whereas several are not. For example, the aerosol hemispheric backscatter fraction and aerosol mass scattering efficiency both depend on the assumed refractive index through Mie theory. Also, the fractional increase in scattering efficiency from hygroscopicity is related to the atmospheric transmittance because both depend on relative humidity. Also, more complete models find substantial aerosol scattering even in areas covered by clouds and would therefore include consideration of the codependence of aerosol scattering in these regions on relative humidity as well.

SOURCE: Penner et al. (1994).

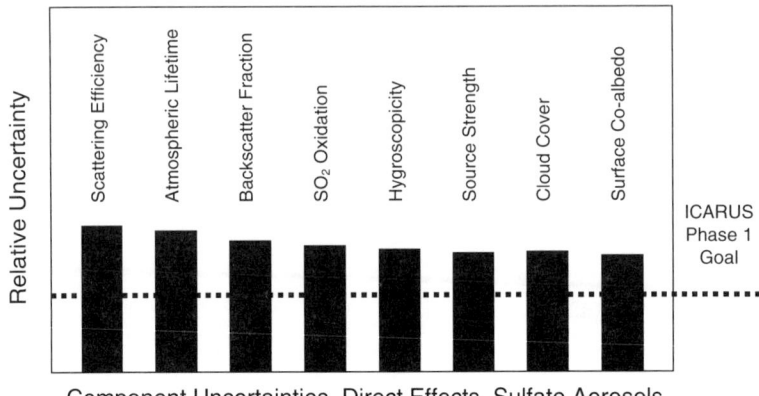

FIGURE 3.6 Plot of the *uncertainties* listed in Table 3.3A for sulfate aerosols, with a qualitative indication of the level to which the uncertainty could be set as a goal for the first phase of ICARUS research.

From Figure 3.6 (showing uncertainties), Figure 3.7 (showing research priorities) can be derived by applying various weighting factors such as those mentioned earlier in this section. In this case, we are suggesting that higher priorities would be given to atmospheric lifetimes (because there is serious doubt that the value used in Table 3.3 is accurate) and to SO_2 oxidation [because this oxidation is critically important also for estimating indirect effects, i.e., defining what fraction of the emitted SO_2 becomes new cloud condensation nuclei (CCN)]. For some other research topics identified in Figure 3.6, a relative decrease in research priority has been suggested, because although the associated uncertainty may be roughly as indicated in Figure 3.5, the necessary data (e.g., on backscattered radiation, cloud cover, and surface albedo) are available, and what is needed is "only" additional analysis of existing data.[3] These comments, however, are meant to be merely qualitative.

In order to define funding priorities (i.e., resource allocations), cost estimates to perform each research task are applied to the research priorities shown in Figure 3.7. The result is shown qualitatively in Figure 3.8, which completes this qualitative description of how only the rightmost bar in Figure 3.4 is derived.

Consider how one of the eight bars shown in Figure 3.8 (funding priorities) would be derived, which in turn provides some details for only one of the four bars in Figure 3.4. In particular, consider details of how to reduce current

---

[3]Whereas data on backscattered radiation are available, few data are available on the backscattered fraction.

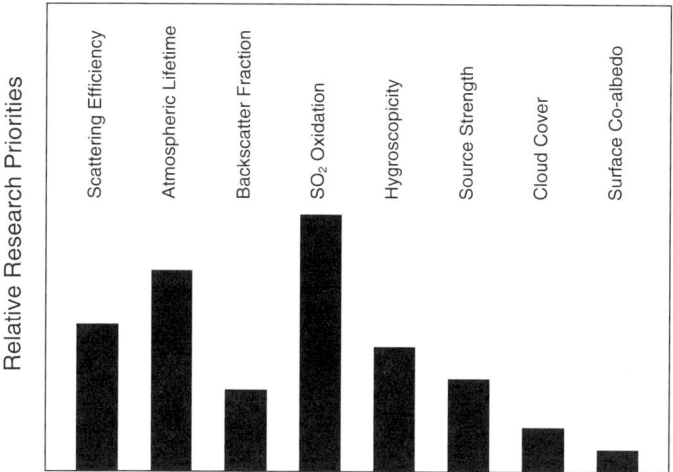

FIGURE 3.7 Qualitative indication of *research priorities* for direct radiative effects of sulfate aerosols, derived from Figure 3.6 (uncertainties) by weighting with such factors as mentioned in the text.

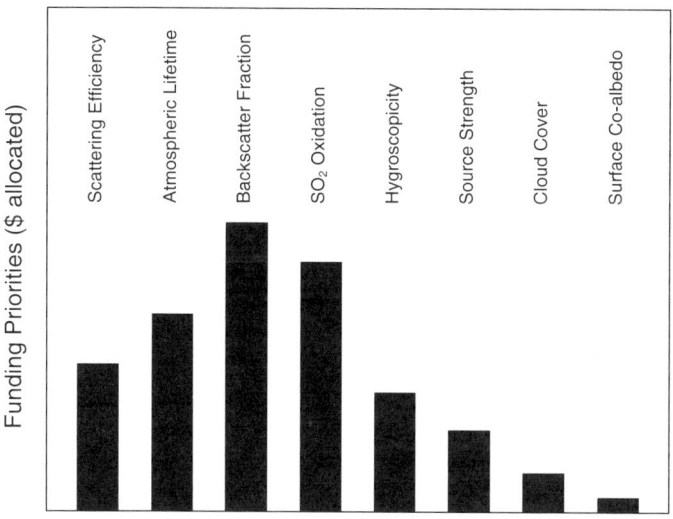

FIGURE 3.8 Qualitative indication of the *relative costs* to reduce the uncertainties shown in Figure 3.6, consistent with the research priorities shown in Figure 3.7, accounting for the cost of performing the research (e.g., a prorated portion of satellite costs to measure backscattered radiation).

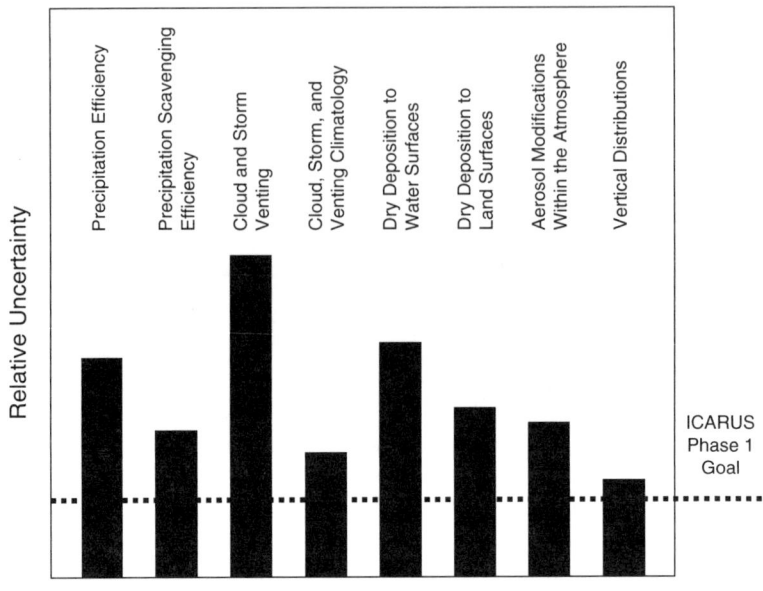

FIGURE 3.9 Qualitative indication of the *relative contributions* from different processes to current uncertainty in the atmospheric lifetime of aerosol sulfate, with a qualitative indication of the level to which the uncertainty could be set as a goal for the first phase of ICARUS research.

uncertainties in the atmospheric lifetime of sulfate aerosols (second bar from the left in Figure 3.8). Figure 3.9 shows (only qualitatively) the relative contributions from different processes to current uncertainties in this lifetime.

Given knowledge about relative uncertainties as shown qualitatively in Figure 3.9, and given still another set of weighting factors (e.g., increase in priority for precipitation efficiencies because they are also a source of major uncertainty in estimates of precipitation within climate models), research priorities can be deduced. Then to these research priorities can be applied the costs of performing the research, leading to funding priorities, as indicated qualitatively in Figure 3.10, which completes this indication of how only one bar in Figure 3.8 could be derived (which in turn suggests the derivation of only one bar in Figure 3.4).

SUMMARY

In this chapter we have outlined a formal framework, based on sensitivity/uncertainty analysis of aerosol-climate models, to select research

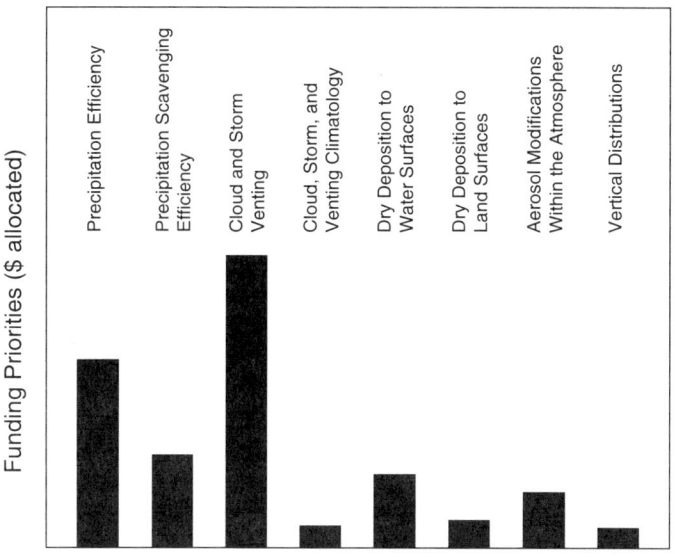

FIGURE 3.10 Qualitative indication of *funding priorities* to reduce the uncertainties shown in Figure 3.9.

priorities. Even in view of this formal procedure, it is clear that, in many respects, it is premature to apply such an analysis to the aerosol-climate problem. Yet, on a qualitative basis it is possible to identify a number of the key research areas that will need to be addressed to reduce uncertainties by using this procedure. Already there are high-priority tasks (Chapter 2) that (we are certain) any future formal sensitivity/uncertainty analysis will demonstrate to be critically important. We therefore conclude that "to get off the ground," ICARUS must plan on some initial "seat-of-the-pants" flying, without the aid of formal sensitivity/uncertainty analysis. The extent to which the ICARUS program incorporates a formal sensitivity/uncertainty analysis approach in setting priorities will have to evolve in time. Chapter 4 presents a possible structure for the ICARUS program and the recommended research.

4

The Proposed ICARUS Program and Recommended Research

It is the panel's judgment that (1) airborne particles can reduce solar radiation reaching the Earth's surface; (2) anthropogenic aerosol provides a negative climate forcing for large regions, and sulfate particles produce forcing in the Northern Hemisphere comparable to that from anthropogenic greenhouse gases (but opposite in sign); and (3) there is substantial uncertainty about the magnitude and spatial distribution of this radiative forcing of the climate by aerosols. Reduction in this uncertainty requires an integrated program that will pull together the research capabilities of the aerosol and atmospheric science communities in this country. The proposed Interagency Climate-Aerosol Radiative Uncertainties and Sensitivities (ICARUS) Program represents such a strategy.

At the outset it should be stressed that the guiding principle of the proposed overall ICARUS research program is integration, within programs sponsored by each agency and among programs sponsored by separate agencies. This latter integration might be termed *horizontal integration*, that is, to promote interagency coordination and cooperation. Integration across modeling, monitoring, process, and laboratory studies is also essential, whether these programs are sponsored by a single agency or more than one agency; this might be called *vertical integration*. The vertical integration of modeling and experimental studies is needed to guide experimental programs via sensitivity analysis and, in turn, to provide data to improve and evaluate models of atmospheric aerosol concentrations, composition, and behavior, including radiative and cloud interactions.

The goal of the ICARUS program is to provide climate models with sufficiently reliable estimates of radiative forcing by airborne particles so that remaining uncertainties in forcing no longer limit quantitative evaluation of climate change. A prime responsibility of the ICARUS organizational structure would be to ensure integration among field and laboratory measurements, model development, and model evaluation.

THE ICARUS STRATEGY

Prior to the present report, many studies have been recommended to reduce uncertainties about direct radiative effects of airborne particles and to permit responsible estimates of their indirect effects. Appendix A gives an incomplete, but illustrative, list of such recommendations, generated during the past quarter century. It is our opinion that currently the greatest need is for administrative leadership in executing a program on aerosols and climate. Thus, we recommend both an administrative strategy and a research strategy, because it is our firm conclusion that developing and applying an effective administrative strategy is equally critical to specifying needed research. To accomplish this goal and achieve the needed integration, the following actions are recommended.

Action 1: Establish Leadership; Empower an Interagency Climate-Aerosol Radiative Uncertainties and Sensitivities (ICARUS) Program

Past failure to undertake needed climate-aerosol research has been caused, mainly, by inadequate coordination, stimulation, recognition, and encouragement. What is needed, as a start, is to stimulate U.S. aerosol and climate scientists (generally affiliated with separate agencies) to pull together as a team. Therefore, as a first step in *Action Item 1,* we recommend that the four federal program managers who requested this report define leadership for U.S. climate-aerosol research by forming a Science Team and an Executive Committee for an ICARUS program.

We expect that these four managers can immediately create this ICARUS program, but, to empower it fully, ICARUS must be organized as a component of the U.S. Global Change Research Program (USGCRP). We therefore recommend that, as soon as possible, ICARUS be recognized as an integral part of the USGCRP. Thereby, no doubt, ICARUS would receive USGCRP guidance for agency representation, Science Team, and Executive Committee membership.

It is appropriate that (1) the ICARUS Executive Committee include government representatives from at least the National Aeronautics and Space Administration (NASA), National Oceanic and Atmospheric Administration

(NOAA), National Science Foundation (NSF), and Department of Energy (DOE), and the director of the USGCRP; (2) the ICARUS Science Team include representatives from the principal climate-aerosol research groups in the United States, as well as selected international researchers; (3) at least annually, a national ICARUS forum be convened for interdisciplinary, interagency, and international discussions about aerosol-climate research progress; and (4) the prime responsibilities of the ICARUS Executive Committee be to set research priorities (see *Action Item 2*) and then coordinate, evaluate, fund, and steer U.S. research on the climate-aerosol problem toward measurable achievements (see *Action Item 3*).

Action 2: Establish Research Priorities; Mobilize ICARUS by Developing Method to Define Research Priorities

The procedure for setting research priorities starts with evaluations of the sensitivities of climate predictions to uncertainties in a host of processes (cloud cover, ocean circulation, air-surface fluxes, snow and ice cover, etc., including changes in atmospheric aerosols); the result is the familiar USGCRP "priority framework."

The USGCRP should reevaluate its science priorities to establish the importance of aerosols to climate sensitivity and uncertainty, relative to all other sources of uncertainties in climate predictions, and then define and adopt an appropriate funding priority for the ICARUS program. It is our expectation that, when this action by the USGCRP has been completed, ICARUS research priorities will appear prominently in the USGCRP's priority framework.

At the present time, the role of clouds in the climate system is listed as the most important research priority for the USGCRP because the degree of response of clouds and water vapor to climate change controls in large measure the overall sensitivity of climate to greenhouse gases. This sensitivity can be measured (and therefore quantified) only by measuring climate change as well as forcing. In view of the importance of knowing the correct forcing for quantifying this process, we find that the quantification of aerosol forcing should be considered of equal priority to the role of clouds in the climate system within USGCRP.

Action 3: Maintain Effective Leadership; Maintain ICARUS's Momentum by Applying Steadily Improving Sensitivity Analyses to Research Designs

In light of the fundamental complexity of the aerosol-climate problem, discerning how to proceed with the ICARUS program will require wise leadership, guided by much-improved mathematical models that are based on accurate and representative field and laboratory data. To advance, there is no scientifically defensible alternative other than to apply the obvious iterative approach: understand the relevant atmospheric processes, improve models, apply sensitivity analyses to design new observational and experimental programs, and use results from these measurement programs to improve models further. Simultaneously, through this iterative process, ICARUS can promote the fundamentally important need to strengthen working relationships between theoreticians and experimentalists—forcing integration between measurement and modeling programs.

Once the USGCRP has defined funding levels for ICARUS and the ICARUS Executive Committee has adopted research priorities, then as *Action Item 3, Phase 1*, the Executive Committee should issue a Request for Proposals (RFP) and thereby invite all interested scientists, federal agencies, and all relevant international research programs to submit detailed plans for proposed climate-aerosol research projects for evaluation by ICARUS Science Team groups. In particular, we encourage NASA and DOE to progress from the available "strategic documents" describing CLIMSAT and an aerosol component of the Atmospheric Radiation Measurement (ARM) Program to "implementation plans," as are now available for the Tropospheric Aerosol Radiative Forcing Observation Experiment (TARFOX) and the North Atlantic Regional Aerosol Characterization Experiment (ACE-2). Thus, in summary, we suggest *Action Item 3, Phase 1*: Issue a single RFP for research plans to be submitted to ICARUS for evaluation.

The next phase is *Action Item 3, Phase 2*: ICARUS Science Team groups should evaluate submitted research plans. We have distinguished this phase in part because conducting reviews is an important, difficult, and time-consuming task; in part because we wanted to mention that ICARUS Science Team groups should perform the reviews; and in part because we felt it appropriate to add the parenthetic remark that we are unconcerned about potential conflicts of interest associated with these reviews. The potential for individual scientists (or groups of scientists) essentially reviewing their own research plans can be avoided by following standard procedures, including requirements for testimonials about any conflicts of interest and appropriate self-regulating disqualification.

An important detail about *Action Item 3* is our recommendation that a number of models be available to address the climate-aerosol issue. No doubt each model will appropriately emphasize different aspects of greatest concern to the developer. Using models to perform sensitivity analyses, each agency should then develop plans for future observational and experimental projects and submit these plans for comment by the ICARUS Science Team. For the next decade or so, models are likely to continue to be, at once, so complicated and yet sufficiently inadequate that derived program plans will almost certainly be model dependent. Consequently, the only procedure available for ICARUS Science Team groups to evaluate project plans independently (if not totally objectively) will be to subject them to analyses by more than one model, developed relatively independently.

The third, continuing phase of maintaining effective leadership is *Action Item 3, Phase 3*: ICARUS Executive Committee should maintain an integrated program. As emphasized earlier, integration must be a cornerstone of an effective program. This integration must occur both horizontally (across agencies and overall projects) and vertically, across the four classes of activities:

1. chemical, climate, and radiative transfer modeling;
2. in situ measurements;
3. remote measurements; and
4. laboratory and field process studies.

Several conclusions emerge from these three proposed action items that should be a central consideration in allocating resources to research on climate forcing by aerosols. These can be listed as a set of questions to be asked regarding any proposed project:

- Is the research part of the integrated interdisciplinary program? Is it part of the systematic approach?
- Is it focused on defining uncertainties and reducing them?
- Is it practical and achievable?
- Does it relate to relevant effects on regional to global scales?

ORGANIZATIONAL STRUCTURE OF THE ICARUS PROGRAM

The suggested organizational structure of the ICARUS program is shown in Figure 4.1. The ICARUS Executive Committee would be composed of the program managers of the four federal agencies and selected members from the ICARUS Science Team. An ICARUS Review Panel, appointed by the USGCR Program Office (PO) and comprised of global change scientists outside the immediate ICARUS program, would periodically review ICARUS and report progress and problems to the USGCRPO.

In the preceding, we have presented a vision of an integrated program to reduce uncertainties about climate forcing by aerosols. We recommend that, to launch ICARUS, the four agency program managers who requested this report (and, as appropriate, their choices of lead ICARUS scientists and representatives from other agencies that commit to funding ICARUS research) proceed to prepare a joint RFP, in coordination with the USGCRP. In the next section, we suggest some broad elements of this first RFP for

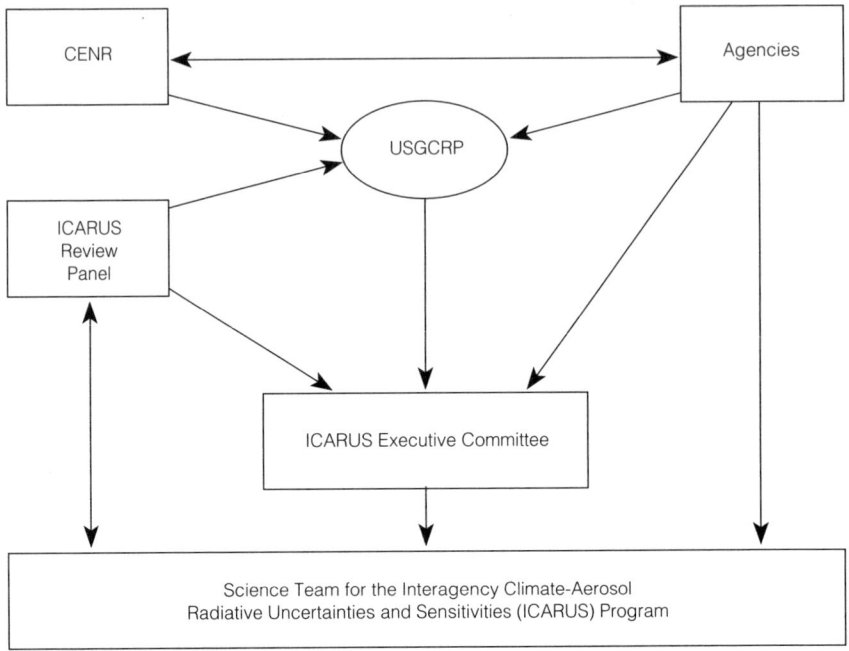

FIGURE 4.1 Organizational structure of the ICARUS program.
NOTE: CENR = Committee on Environment and Natural Resources.

ICARUS. When general features of this joint RFP are agreed upon (including identification of lead agencies for its various components, and after proposals have been reviewed by a panel created by the ICARUS Executive Committee), contracts should be let by each agency for its identified ICARUS components, according to each agency's standard practices.

RESEARCH PROGRAM

In Chapter 2 we identified the components of research required to increase our understanding of aerosol forcing of climate. We stressed that this should be an integrated program involving process studies, model development and evaluation, field measurements, and technology development. In this final section of the report, we recommend specific research tasks, together with estimated costs and durations.

From the point of view of cost, four categories of research projects emerge from the analysis in Chapter 2:

Category of Project	Cost/Project (million $)
Process research (laboratory, modeling, small-scale field studies) and technology development (excluding satellite development)	<1
System integration and assessment (model development, sensitivity analysis)	1-5
Large, multiplatform field studies (e.g., ACE-1)[1] and global surface-based monitoring network	10-20
Satellite system development	>50

---

[1] An example of the difficulty of estimating the cost of a large, multiplatform field program is provided by the Southern Hemisphere Marine Aerosol Characterization Experiment (ACE-1) (B. Huebert, private communication, 1995). A total of $8 million to $9 million in proposals was submitted to NSF and NOAA for support of individual Principal Investigators (PIs); about $4 million was funded. Several individuals who have funding from NASA's aerosol program are planning to participate in ACE-1. If three years' duration is estimated for each project, at $150,000 per year for each of five PIs, the total is approximately $2 million. Others already have NSF or NOAA funding; it is probably reasonable to assume another $1 million from existing grants. The estimated NSF Observing Facilities Deployment Pool cost for C-130 flight hours and integrated sounding systems (ISS) is $1.5 million, that is just the "out-of-pocket costs" of operating the aircraft and having the Research Aviation Facility personnel in the field. If depreciation of the plane, salaries for pilots and technicians, etc., were included, one could easily double the amount. Modifications to the plane alone will cost several million dollars, but that should be amortized over multiple years and projects. It is probably reasonable to assume a total of $3 million for the total cost of the plane. The NOAA ship will cost about $20,000 per day for out-of-pocket costs. For a 60-day deployment (not including the return trip), the total is $1.2 million. Again, that amount does not cover salaries and the total cost of operating the ship. A more conservative estimate is $2 million. How can one evaluate the contribution of salaries to an effort of this type? Over the three years of planning and a few years of analysis afterward, perhaps 100 person-years of non-grant salary and benefits (at an average of $100,000 per year with overhead) might go into ACE-1. That is $10 million, but can one say it is an extra cost of doing ACE-1, since those people would presumably be employed anyway? A total of $2 million for all non-U.S. contributions is probably reasonable. Finally, how does one count the cost of satellite data? The experiment will rely heavily on Sea-Viewing Wide Field of View Sensor and Advanced Very High Resolution Radiometer data. What fraction of that cost should ACE-1 assume?

Clearly, the total cost is a strong function of the assumptions made about what to count. One could conclude that the incremental cost of doing ACE-1 will be less than $10 million, or one could equally honestly come up with real costs in the vicinity of $30 million, over a two- to three-year period. The above estimates do not include the costs associated with constructing and maintaining the facilities to be used, including those at Cape Grim and the ISS systems.

Similar, but even broader, problems arise if an attempt is made to estimate the financial support to an entire area of research activity—funding for atmospheric aerosol radiative

Each category represents an important component of the overall plan, yet the panel recognizes that the manner in which funding decisions are made will likely vary according to the category of project. In spite of the substantial cost of new satellite systems for remote sensing, this type of development is likely to be undertaken by a single agency such as NASA. On the other hand, large, multiplatform field studies are generally funded by a consortium of agencies. Individual process studies are normally funded by a single agency. The goal of the ICARUS management structure is to achieve communication among the different agencies supporting the research, so that the various projects constitute an integrated attack on the problem of aerosol forcing of climate.

It is important to note that the costs itemized above and below assume the existence of an atmospheric science research infrastructure. This includes active research groups in universities and the national laboratories, ships (the University-National Oceanographic Laboratory System and NOAA fleets), and aircraft (e.g., those at the National Center for Atmospheric Research and NASA) that are configured to conduct research of this type, and supercomputing facilities that can run the complex models. If budgetary constraints were to necessitate a substantial reduction in the current infrastructure, it would become much more costly to redevelop the capability for answering these questions.

Global Climate Model Development

Project 1. Sensitivity and Uncertainty of Aerosol Forcing in Global Climate Models

Using existing global climate and global chemistry models, determine relative sensitivities of both global and regional climate predictions to uncertainties in aerosol forcing versus uncertainties in other factors, such as cloud parameterizations. Using global chemistry models examine the sensitivities of aerosol forcing predictions [at scales from global circulation model (GCM)

---

forcing research as an example. An estimate of federal funding for these areas was made by sending a questionnaire to all agencies supporting aerosol research or aerosol forcing research. Since there are no rules, standards, or requirements for formally reporting such information, that received is uneven in content. However, it does furnish some idea of the size of the total federal aerosol research effort and the fraction of that devoted to aerosol forcing research.

The survey showed that aerosol research is supported by nine agencies, although only four—DOE, NASA, NOAA, and NSF—directly support aerosol forcing research. The figures supplied by the agencies show that in FY 1994 almost $30 million was dedicated to aerosol research with about 25 percent of that, at most, supporting research that the four agencies consider to be driven primarily by the climate forcing issue.

grid scales to the global scale] to uncertainties in microphysical properties (aerosol size, aerosol composition, relative humidity) and processes (sulfate production pathways and rates, and vertical convective transport). We recommend up to four such projects to enable assessment of the influence of model platform and assumptions employed in individual models.

Number of projects	4
Duration of each project	2 years
Cost per project	$500,000 per year

Project 2. Development of Global-Scale Aerosol Radiative Forcing Models

Develop and evaluate the next-generation models of aerosol radiative forcing at scales from regional to global, incorporating all chemical, physical, and meteorological processes that are important in determining the concentrations, radiative and cloud-nucleating properties, and direct and indirect radiative forcing from natural and anthropogenic aerosols. We recommend two such projects, employing different GCM and global chemical model platforms. Note that aerosol submodel development will require integration with research on aerosol process models.

Number of projects	2
Duration of each project	4 years
Cost per project	$500,000 per year

Process Research

Project 1. Aerosol Formation and Growth by Nucleation and Gas-to-Particle Conversion

The goal is to ascertain to what extent homogeneous nucleation is occurring in the atmosphere, where (e.g., marine boundary layer, free troposphere) it is occurring, under what conditions, what chemical species are involved, and to what extent theoretical treatments of nucleation can represent observed new particle formation in the atmosphere. This project will consist of several individual, highly integrated, projects, including

1a. elucidation of the linkage between dimethyl sulfide (DMS) SO_2 emissions and nss (non-sea salt)-sulfate formation in the marine and continental boundary layer based on evaluation of boundary-layer aerosol data;

1b. evaluation of nucleation as a source of new particles in the free troposphere based on evaluation of free tropospheric aerosol data;

1c. evaluation of the applicability of nucleation theory to describe the atmospherically relevant systems of H_2SO_4-H_2O, NH_3-H_2SO_4-H_2O;

1d. evaluation, through laboratory experiments, of the conversion of low vapor pressure organic species to aerosols through nucleation processes; and

1e. elucidation of tropospheric aerosol growth processes involving gas-to-particle conversion both in clear air and in-cloud, based on analysis of ambient size distribution data and chemical routes of gas-to-particle conversion.

Note that each of the projects must be closely integrated with field measurement programs.

Number of projects	8
Duration of each project	3 years
Cost per project	$250,000 per year

Project 2. Aerosol and Cloud Optical Properties

The goal is to determine the extent to which the theoretical treatments of aerosol and cloud optical properties that are used in aerosol forcing models are applicable to ambient aerosols and clouds. This includes laboratory experiments on pure and mixed-component aerosols, together with in situ measurements of ambient particles for important classes of tropospheric aerosols. Also included are theoretical treatments and in situ and satellite measurements of cloud optical properties and albedo to determine the accuracy of the treatment of cloud optical properties in global models.

Number of projects	4
Duration of each project	3 years
Cost per project	2 at $250,000 per year
	2 at $500,000 per year

Project 3. The Aerosol-CCN-CDNC-Cloud Albedo Linkage

For fundamental evaluation of the effect of anthropogenic emission changes on cloud optical properties (the indirect effect), it is necessary to develop a first-principles understanding of the relation between changes in aerosol number concentration (as a function of aerosol molecular composition) and cloud condensation nuclei (CCN) behavior; between CCN and resulting cloud drop number concentrations (CDNC); and between CDNC and cloud albedo. This project includes detailed aerosol process modeling, coupled with cloud microphysical modeling. It also includes laboratory and field studies of the CCN properties of aerosols in both clean and anthropogenically influenced regions of the atmosphere. Note the integration with technological development of new CCN spectrometers. Aerosol model parameterizations of the aerosol-CCN-CDNC-cloud albedo linkage should be developed for GCMs and ACTMs (atmospheric chemical transport models).

Number of projects	4
Duration of each project	4 years
Cost per project	2 at $250,000 per year
	2 at $500,000 per year

Project 4. Aerosol Sinks

The prime goal of the dry deposition studies is to obtain reliable, particle size-specific data in the field (i.e., not just from wind tunnels) for dry deposition to "real-world" collectors—from forests in inhomogeneous terrain to the oceans under a variety of conditions. To achieve the essential goal of obtaining particle size-specific dry deposition velocities, developments in technology (e.g., for eddy-flux measurements) and techniques (e.g., for measuring monodisperse particles actually deposited) almost certainly will be necessary. The prime emphasis of the precipitation scavenging field studies will be to develop parameterizations for storm venting for use in global-scale models for all relevant species (especially for particles as a function of their sizes), for all climatologically important cloud and storm types, and for representative ranges of pollution loadings and storm microphysical and dynamical variables. Analysis of field data must be performed with appropriate mesoscale models of precipitation formation, efficiencies, and scavenging.

Number of projects	6
Duration of each project	3 years
Cost per project	$500,000 per year

Multiplatform Field Campaigns

Project 1. Multiplatform Field Campaign

One multiplatform field campaign will take place every two years for the next decade. The goals are to integrate satellite radiation measurements and surface-based column-integrated radiation measurements with in situ (aircraft) chemical, physical, and optical measurements in both clear-sky and cloud environments, and in both clean and anthropogenically influenced air masses. In situ measurement of CCN, CDNC, aerosol size and composition distribution, optical depth, and cloud albedo will be needed. The goal is to perform closure studies to assess the accuracy of treatment of aerosol and cloud radiative processes in GCMs and ACTMs.

Number of projects	5 in 10 years
Duration of each project	3 years
Cost per project	$10 million

Project 2. Mobile Platforms

Conduct a research measurement effort involving regularly scheduled flights by a suitably instrumented aircraft in a nearly continuous circuit over selected surface aerosol monitoring sites. A similar routine measurement program from ships of opportunity is also suggested as is a program of balloon-borne measurements. All of these observations are needed to provide ground truth for satellite observations and the information necessary to determine chemical composition for source identification and quantification.

Number of projects	1 series of continuing observations by aircraft, ship, and balloons
Duration of project	3-5 years (with significantly lower level of activity thereafter)
Cost per project	$1.5 million per year for aircraft $250,000 per year for ships $250,000 per year for balloons

Project 3. Surface-Based Stations for Continuous Monitoring of Aerosols

Establish a dual-density network of surface-based stations for continuous monitoring of aerosols. A high-density network of about 30 stations spanning North America will characterize the spatial distribution, seasonal variability, and trends of aerosol optical depth with a spatial coverage suitable for testing chemical transport and radiative transfer models. A companion network of about 10 stations will provide detailed information on means, variability, and trends of key aerosol radiative, chemical, and microphysical properties, for different aerosol types, that are used in chemical transport and radiative transport models. These networks will be supplemented with a systematic program of ship- and airborne surveys to characterize the horizontal and vertical distributions, over the global oceans and North America, of the same aerosol properties studied in the low-density network.

Description	Number of Projects	Duration of Each Project (years)	Cost per Project per Year ($)	Total Cost over 10 Years (million $)
Surface monitoring—aerosol optical depth	30	10	10,000	3
Surface monitoring—aerosol properties (including balloon-borne monitoring of stratospheric aerosol size distribution)	8	10	250,000	20
Systematic airborne surveys	1	3	1 million	3
Systematic shipborne surveys	2	5	250,000	2.5

Satellite System Development

Recommended Satellite Observations

It is recommended that data from current (e.g., Advanced Very High Resolution Radiometer) and future [e.g., Multiangle Imaging Spectroradiometer (MISR)] satellite instruments be integrated with laboratory, theoretical and in situ optical, chemical, and microphysical studies in order to quantify the uncertainty inherent in passive radiometry and optimize the information provided. For improving retrievals of tropospheric aerosol optical depth, especially over land, and for determining the vertical distribution of tropospheric aerosols globally, we recommend that a spaceborne aerosol backscatter measuring lidar operating at a minimum of 2 wavelengths with the capability of measuring depolarization be flown as soon as possible. On the same spacecraft, a carefully designed set of ancillary instruments boresighted with the lidar should be flown, e.g., an imager to aid in horizontal extrapolation. It is necessary to have in place a network of supporting ground-based and airborne in situ measurements to constrain the remotely sensed retrievals of optical depth to meet the required measurement accuracies for both tropospheric and stratospheric aerosols. Further, we recommend that the Stratospheric Aerosol and Gas Experiment (SAGE) series of satellite remote sensors be continued for the measurement of aerosol optical depth in the stratosphere, and that global coverage be achieved. Presently, NASA's Earth Observing System (EOS) program is developing SAGE III instruments for flights beginning in 1998. However, global coverage requires a spacecraft in both polar and mid-inclined orbits, simultaneously.

Non-Satellite Technology Development

The goal is to advance the state of the art of instrumentation needed to assess aerosol forcing. Particular focus is on the development of compact and lightweight sensors for aircraft use.

Project 1. Development of Advanced Aerosol Optical Instrumentation

  1a. In situ measurement of aerosol light absorption and refraction index
  1b. Continuous polar nephelometer

Number of projects   2
Duration of each project   3 years
Cost per project $250,000 per year

Project 2. Development of a Compact CCN Spectrometer for Aircraft Use

Number of projects 1
Duration of each project 3 years
Cost per project $250,000 per year

Project 3. Development of Rapid, Lightweight Samplers for In Situ Chemical Analysis of Atmospheric Aerosols

Number of projects 2
Duration of each project 4 years
Cost per project $250,000 per year

Satellite Technology Development

Project 1. Satellite Development

The goal is to develop the capability to monitor extinction profiles (i.e., vertically resolved aerosol optical depths) on a global scale from an Earth-orbiting spacecraft. Lidars appear to be the strongest candidates for producing unambiguous vertically resolved measurements in the troposphere with the accuracies needed, especially when coupled to local in situ measurements that constrain the inverted solutions. In the stratosphere, where optical depths are low and aerosols more homogeneous, limb occultation passive instruments provide the necessary aerosol information. All passive sensor data, however, especially those from NASA's EOS program, which begins its flights in 1998 (SAGE III, Moderate-Resolution Imaging Spectroradiometer, MISR, etc.), will be integrated into this program.

Number of projects 1
Duration of each project 5 years
Cost per project $50 million

We recognize that this integrated program of research on climate forcing by aerosols will require the commitment of resources for a sustained period of time. We also recognize that acquiring new resources during a period of fiscal constraint is likely to be difficult. We urge the agencies that are involved to recognize the strength of the evidence for the existence of a substantial aerosol forcing and base the priority for this proposed research accordingly.

System Integration and Assessment

The foregoing projects will be useful for reducing uncertainties in climate forcing by aerosol only if three further activities are undertaken. It is essential that the ICARUS management structure formally organizes projects on coordination, integration, and assessment.

Project 1. Coordination

The goal is to maintain communication among projects regarding technical details of the ongoing efforts. Such coordination is necessary, for example to ensure that models produce results with averaging times that are relevant to in situ measurements. Ground-based, aircraft, and satellite observations must coincide in space and time if closure exercises are to be conducted. Two specific activities are envisaged:

1. a small coordination project that actively tracks all the projects and seeks to develop logical interfaces between them ($100,000 per year), and

2. annual Science Team meetings for reporting research results and exchanging data ($100,000 per year).

Project 2. Integration and Assessment

The final product of all of the projects is supposed to be information on climate forcing by aerosols that is tailored for use in studies of climate response. Thus, there is a need for active interaction with the meteorological and climate modeling community as well as the meteorological data analysis community. In addition, there is a need for delivery of these research results to the Intergovernmental Panel on Climate Change (IPCC). Presently, U.S. contributions to IPCC regarding aerosols are arranged on an ad hoc basis, and there is no means for coordinating U.S. scientific input other than by actions of individual scientists.

The needs here are for time and travel for representatives of ICARUS to deliver results to the user community ($50,000 per year).

Project 3. Integration of U.S. Research on Aerosol Forcing of Climate

The dominant role of ICARUS will be to integrate U.S. research on aerosol forcing of climate. Some of this integration will be funded indirectly, through each agency's representatives on ICARUS's Executive Committee and through sponsorship of principal investigators who will be members of the ICARUS' Science Team. Other integration tasks, however, will require separate and sometimes substantial funding, including establishing and maintaining quality assured data bases; providing coordination and support services for field campaigns; funding model and measurement intercomparison studies; and performing a variety of management tasks, such as hosting workshops and annual meetings, and publishing proceedings. Consequently, ICARUS's first RFP should solicit proposals for providing integration services, possibly through a management contract.

Number of projects	1
Duration of each project	Continuing
Cost for the single project	$1 million per year

References

Ackerman, A.S., O.B. Toon, and P.V. Hobbs, 1995. A model for particle microphysics, radiative transfer, and turbulent mixing in the stratocumulus topped marine boundary layer: Model description and comparison with observations. *J. Atmos. Sci.*, in press.

Alkezweeny, A.J., A.D. Burrows, and A.C. Grainger, 1993. Measurements of cloud-droplet size distributions in polluted and unpolluted stratiform clouds. *J. Appl. Meteorol. 32*, 106-115.

Allen, A.G., R.M. Harrison, and K.W. Nicholson, 1991. Dry deposition of fine aerosol to a short grass surface. *Atmos. Environ. 25A*, 2671-2676.

Andreae, M.O., 1995. Climatic effects of changing atmospheric aerosol levels. In *World Survey of Climatology,* Vol. 16: *Future Climates of the World.* A. Henderson-Sellers, ed., Elsevier, Amsterdam.

Ansmann, A., M. Riebesell, and C. Weitkamp, 1990. Measurement of atmospheric aerosol extinction profiles with a raman lidar. *Opt. Lett. 15*, 746-748.

Arny, D.C., S.E. Lindow, and C.D. Upper, 1976. Frost sensitivity of Zea Mays increased by application of *Pseudomonas syringai*. *Nature 262*, 282-284.

Arrhenius, S., 1896. On the influence of carbonic acid in the air upon the temperature of the ground. *Phil. Mag. 41*, 237-276.

Atherton, C.S., 1994. Predicting tropospheric ozone and hydroxyl radical in a global three-dimensional, chemistry, transport, and deposition model. Ph.D. dissertation, University of California, Davis.

Ball, R.J., and G.P. Robinson, 1982. The origin of haze in the central United States and its effect on solar radiation. *J. Appl. Meteorol. 21*, 171-188.

Bates, T., J. Gras, and B. Huebert (eds.), 1993. *ACE-1, 1993. Radiative Forcing Due to Aerosols in the Remote Marine Atmosphere—Science and Implementation Plan.* Available from T. Bates, NOAA/PMEL, 7600 Sandpoint Way NE, Seattle, WA 98115 (bates@pmel.noaa.gov).

Baumgardner, D., J.E. Dye, B. Gandrud, D. Rogers, K. Weaver, R.G. Knollenberg, R. Newton, and R. Gallant, 1995. The multiangle aerosol spectrometer probe: A new instrument for airborne particle research. *Proceedings Ninth Symposium on Meteorological Observations and Instrumentation*, American Meteorological Society, Charlotte, N.C., March 27-31.

Benkovitz, C.M., C.M. Berkowitz, R.C. Easter, S. Nemesure, R. Wagener, and S.E. Schwartz, 1994. Sulfate over the North Atlantic and adjacent continental regions: Evaluation for October and November 1986 using a three-dimensional model driven by observation-derived meteorology. *J. Geophys. Res. 99*, 20725-20756.

Berresheim H., F.L. Eisele, D.J. Tanner, L.M. McInnes, D.C. Ramsey-Bell, and D.S. Covert, 1993. Atmospheric sulfur chemistry and cloud condensation nuclei (CCN) concentrations over the northeastern Pacific coast. *J. Geophys. Res. 98*, 12701-12711.

Bigg, E.K., 1973. Ice nucleus measurements in remote areas. *J. Atmos. Sci. 30*, 1153-1157.

Boers, R., G.P. Ayers, and G.L. Gras, 1994. Coherence between seasonal variation in satellite-derived cloud optical depth and boundary layer CCN concentration at a mid-latitude Southern Hemisphere station. *Tellus 46B*, 123-131.

Bohren, C.F., and D.R. Huffman, 1983. *Absorption and Scattering of Light by Small Particles*. John Wiley, New York.

Bolin, B., and R.J. Charlson, 1976. On the role of the tropospheric sulfur cycle in the shortwave radiative climate of the Earth. *Ambio 5*, 47-54.

Boucher, O., 1995. GCM estimate of the indirect aerosol forcing using satellite-retrieved cloud droplet effective radii. *J. Climate 8*, 1403-1409.

Boucher, O., and T.L. Anderson, 1995. GCM assessment of the sensitivity of direct climate forcing by anthropogenic sulfate aerosols to aerosol size and chemistry. *J. Geophys. Res.*, in press.

Boucher, O., and V. Lohmann, 1995. The sulfate-CCN-cloud albedo effect: A sensitivity study with two general circulation models. *Tellus 47B*, 281-300.

Boucher, O., and H. Rodhe, 1994. *The Sulfate-CCN-Cloud Albedo Effect: A Sensitivity Study*. Report CM-83, Department of Meteorology, International Institute of Meteorology, Stockholm University.

Braham, R.R., Jr., and P.A. Spyers-Duran, 1974. Ice nucleus measurements in an urban atmosphere. *J. Appl. Meteorol. 13*, 940-945.

Cambray, R.S., P.A. Cawse, J.A. Garland, J.A.B. Gibson, P. Johnson, G.N.J. Lewis, D. Newton, L. Salmon, and B.O. Wade, 1987. Observations on radioactivity from the Chernobyl accident. *Nuclear Energy 26*, 77-110.

Chamberlain, A.C., 1991. *Radioactive Aerosols*. Cambridge University Press, New York, 255 pp.

Charlson, R.J., 1974. Sulfuric acid ammonium sulfate aerosol—Optical detection in St. Louis Region. *Science 184*, 156-158.

Charlson, R.J., and J. Heintzenberg (eds.), 1995. Aerosol forcing of climate. In *Report of the Dahlem Workshop on Aerosol Forcing*, Berlin, April 24-29. Wiley, Chichester, U.K.

Charlson, R.J., J.E. Lovelock, M.O. Andreae, and S.G. Warren, 1987. Oceanic phytoplankton, atmospheric sulfur, cloud albedo and climate. *Nature 326*, 655-661.

Charlson, R.J., J. Langner, and H. Rodhe, 1990. Sulphate aerosol and climate. *Nature 348*, 22.

Charlson, R.J., J. Langner, H. Rodhe, C.B. Leovy, and S.G. Warren, 1991. Perturbation of the Northern Hemispheric radiative balance by backscattering from anthropogenic sulfate aerosols. *Tellus 43AB*, 152-163.

Charlson, R.J., S.E. Schwartz, J.M. Hales, R.D. Cess, J.A. Coakley, Jr., J.E. Hansen, and D.J. Hofmann, 1992a. Climate forcing by anthropogenic aerosols. *Science 255*, 423-430.

Charlson, R.J., T.L. Anderson, and R.E. McDuff, 1992b. The sulfur cycle. In *Global Biogeochemical Cycles*, S.S. Butcher et al. (eds.). Academic Press, London.

Chuang, C.C., and J.E. Penner, 1995. Effects of aerosol sulfate on cloud drop nucleation and optical properties. *Tellus*, in press.

Chuang, C.C., J.E. Penner, K.E. Taylor, and J.J. Walton, 1994. Climate effects of anthropogenic sulfate: Simulations from a coupled chemistry/climate model. Pp. 170-174 in *Proceedings of the Conference on Atmospheric Chemistry*, American Meteorological Society, Nashville, Tenn., January 23-28.

Chylek, P., and J.A. Coakley, 1974. Aerosols and climate. *Science 183*, 75-77.

Clarke, A.D., 1993. Atmospheric nuclei in the Pacific midtroposphere: Their nature, concentration, and evolution. *J. Geophys. Res. 98*, 20633-20647.

Coakley, J.A., Jr., R.D. Cess, and F.B. Yurevich, 1983. The effect of tropospheric aerosols on the Earth's radiation budget: A parameterization for climate models. *J. Atmos. Sci. 40*, 116-138.

Coakley, J.A., Jr., R.L. Bernstein, and P.A. Durkee, 1987. Effect of ship-track effluents on cloud reflectivity. *Science 273*, 1020-1022.

Cooke, W.F., and J.J.N. Wilson, 1995. A global black carbon aerosol model. *J. Geophys. Res.* (submitted).

Cooper, W.A., R.T. Bruintjes, and G.K. Mather, 1995. Some calculations pertaining to hygroscopic seeding with flares. *Proceeding of the American Meteorological Society Conference on Cloud Physics*, Dallas, Tex., January 15-20.

Crutzen, P.J., and P.H. Zimmermann, 1991. The changing photochemistry of the troposphere. *Tellus 43*, 136-151.

d'Almeida, G., P. Koepke, and E. Shettle, 1991. *Atmospheric Aerosols: Global Climatology and Radiative Characteristics*. A. Deepak Publishing, Hampton, Va., 561 pp.

DeMott, P.J., M.P. Meyers, and W.R. Cotton, 1994. Parameterization and impact of ice initiation processes relevant to numerical model simulation of cirrus clouds. *J. Atmos. Sci. 51*, 77-90.

Dick, W.D., P.H. McMurry, and J.R. Bottiger, 1994. Size- and composition-dependent response of the DAWN-A multiangle single-particle optical detector. *Aerosol Sci. Tech. 20*, 345-362.

Dignon, J., and S. Hameed, 1989. Global emissions of nitrogen and sulfur oxides from 1860 to 1980. *J. Air Pollution Control Assoc. 39*, 180-186.

Durkee, P.A., 1988. Observations of aerosol cloud interaction in satellite-detected visible and near-infrared radiance. Pp. 157-160 in *Preprints, Symposium on Role of Clouds in Atmospheric Chemistry and Global Climate*. American Meteorological Society, Boston.

Durkee, P.A., F. Pfeil, E. Frost, and R. Shema, 1991. Global analysis of aerosol particle characteristics. *Atmos. Environ. 25A*, 2457-2471.

Dutton, E.G., and J.R. Christy, 1992. Solar radiative forcing at selected locations and evidence of global lower tropospheric cooling following the eruptions of El Chichon and Pinatubo. *Geophys. Res. Lett. 19*, 2313-2316.

Engardt, M., and H. Rodhe, 1993. A comparison between patterns of temperature trends and sulfate aerosol pollution. *Geophys. Res. Lett. 20*, 117-120.

Fitzgerald, J.W., and P.A. Spyers-Duran, 1973. Changes in cloud nucleus concentration and cloud droplet size distribution associated with pollution from St. Louis. *J. Appl. Meteorol. 12*, 511-516.

Flossmann, A.I., and H.R. Pruppacher, 1988. A theoretical study of the wet removal of atmospheric pollutants. Part III: The uptake, redistribution, and deposition of $(NH_4)_2SO_4$ particles by a convective cloud using a two-dimensional cloud dynamics model. *J. Atmos. Sci. 43*, 1857-1871.

Flowers, E.C., R.A. McCormick, and K.R. Kurfis, 1969. Atmospheric turbidity over the United States, 1961-1966. *J. Appl. Meteorol. 8*, 955-962.

Frisbie, P.R., and J.G. Hudson, 1993. Urban cloud condensation nuclei spectral flux. *J. Appl. Meteorol. 32*, 666-676.

Fukuta, N., and V.K. Saxena, 1979a. The principle of a new horizontal thermal gradient cloud condensation nucleus counter. *J. Res. Atmos. 13*, 169-188.

Fukuta, N., and V.K. Saxena, 1979b. A horizontal thermal gradient cloud chamber, *J. Appl. Meteorol. 18*, 1352-1362.
Galbally, I.E. (ed.), 1989. *The International Global Atmospheric Chemistry (IGAC) Programme*. Commission on Atmospheric Chemistry and Global Pollution of the International Association of Meteorology and Atmospheric Physics, ISBN 0 643 05062 0, available from the IGAC Project Office, Massachusetts Institute of Technology, 54-1824, Cambridge, MA 02139.
Galloway, J.N., J.E. Penner, C.S. Atherton, D.R. Hastie, J.M. Prospero, H. Rodhe, R.S. Artz, Y.J. Balkanski, H.G. Bingemer, R.A. Brost, S. Burgermeister, G.R. Carmichael, J.S. Chang, R.J. Charlson, S. Cober, W.G. Ellis, Jr., C.F. Fischer, J.M. Hales, T. Iversen, D.J. Jacob, K. John, J.E. Johnson, P.S. Kasibhatla, J. Langner, J. Lelieveld, H. Levy II, F. Lipschutz, J.T. Merrill, A.F. Michaels, J.M. Miller, J.L. Moody, J. Pinto, A.A.P. Pzenny, P.A. Spiro, L. Tarrason, S.M. Turner, and D.M. Whelpdale, 1992. Sulfur and nitrogen levels in the North Atlantic Ocean's atmosphere: A synthesis of field and modeling results. *Global Biogeochemical Cycles 6*, 77-100.
Garland, J.A., and L.C. Cox, 1982. Deposition of small particles to grass. *Atmos. Environ. 16*, 2699-2702.
Garrett, T.J., and P.V. Hobbs, 1995. Long-range transport of continental aerosols over the Atlantic Ocean and their effects on cloud structures. *J. Atmos. Sci. 52*, 2977-2984.
Ghan, S.J., C.C. Chuang, and J.E. Penner, 1993. A parameterization of cloud droplet nucleation. Part I: Single aerosol types. *Atmos. Res. 30*, 198-221.
Gidel, L.T., 1983. Cumulus cloud transport of transient tracers. *J. Geophys. Res. 88*, 6587-6599.
Gillani, N.V., S.E. Schwartz, W.R. Leaitch, J.W. Strapp, and G.A. Isaac, 1995. Field observations in continental stratiform clouds: Partitioning of cloud particles between droplets and unactivated interstitial aerosols. *J. Geophys. Res. 100*, 18687-18706.
Giorgi, F., 1988. Dry deposition velocities of atmospheric aerosols as inferred by applying a particle dry deposition parameterization to a general circulation model. *Tellus 40B*, 23-41.
Graf, H.-F., I. Kirchner, A. Robock, and I. Schult, 1993. Pinatubo eruption winter climate effects: Models versus observations. *Climate Dynamics 9*, 81-93.
Grassl, H., 1988. What are the radiative and climatic consequences of the changing concentration of atmospheric aerosol particles? In Proceedings of the Dahlem Workshop, *The Changing Atmosphere*, F.S. Rowland and I.S.A. Isaksen (eds.). John Wiley & Sons, Chichester.
Grovenstein, J.D., K.V. Saxena, and A.P. Durkee, 1994. Impact of anthropogenic and natural aerosols on cloud albedo. *Eos Trans. Geophys. Union 75*, 73.
Hagen, D.E., J. Schmidt, M. Trueblood, J. Carstens, D.R. White, and D.J. Alofs, 1989. Condensation coefficient for water in the UMR simulation chamber. *J. Atmos. Sci. 46*, 803-816.
Han, Q., W.B. Rossow, and A.A. Lacis, 1994. Near-global survey of effective droplet radii in liquid water clouds using ISCCP data. *J. Climate 7*, 465-497.
Hansen, J.E., and A.A. Lacis, 1990. Sun and dust versus greenhouse gases: An assessment of their relative roles in global climate change. *Nature 346*, 713-719.
Hansen, J.E., A. Lacis, R. Ruedy, and M. Sato, 1992. Potential climate impact of Mt. Pinatubo eruption. *Geophys. Res. Lett. 19*, 215-218.
Hansen, J.E., A. Lacis, R. Ruedy, M. Sato, and H. Wilson, 1993a. How sensitive is the world's climate? *National Geographic Research and Exploration 9*, 142-158.
Hansen, J., W. Rossow, and I. Fung (eds.), 1993b. *Long-Term Monitoring of Global Climate Forcings and Feedbacks*. NASA Conference Publication 3234, available from NASA Goddard Space Flight Center, Greenbelt, Md. 20771.
Hansen, J.E., M. Sato, and R. Ruedy, 1995. Long-term changes of the diurnal temperature cycle: Implications about mechanisms of global climate change. *Atmos. Res.*, in press.

Harshvardhan, 1979. Perturbation of the zonal radiation balance by a stratospheric aerosol layer. *J. Atmos. Sci. 36,* 1274-1285.

Hayasaka, T., T. Nakajima, S. Ohta, and M. Tanaka, 1992. Optical and chemical properties of urban aerosols in Sendai and Sapporo, Japan. *Atmos. Environ. 26A,* 2055-2062.

Haywood, J.M., and K.P. Shine, 1995. The effect of anthropogenic sulfate and soot aerosol on the clear sky planetary radiation budget. *Geophys. Res. Lett.,* in press.

Hegg, D., 1994. Cloud condensation nucleus-sulfate mass relationship and cloud albedo. *J. Geophys. Res. 99,* 25903-25907.

Hegg, D.A., R.J. Ferek, and P.V. Hobbs, 1993. Light scattering and cloud condensation nucleus activity of sulfate aerosol measured over the Northeast Atlantic Ocean. *J. Geophys. Res. 98,* 14887-14894.

Hervig, M.E., J.M. Russell III, L.L. Gordley, J.H. Park, and S.R. Drayson, 1993. Observations of aerosol by the HALOE experiment onboard UARS: A preliminary validation. *Geophys. Res. Lett. 20,* 1291-1294.

Heymsfield, A.J., and R.M. Sabin, 1989. Cirrus crystal nucleation by homogeneous freezing of solution droplets. *J. Atmos. Sci. 46,* 2252-2264.

Hobbs, P.V., L.J. Radke, and S.E. Shumway, 1970. Cloud condensation nuclei from industrial sources and their apparent influence on precipitation in Washington State. *J. Atmos. Sci. 27,* 81-89.

Hobbs, P.V., G.C. Bluhm, and T. Ohtake, 1971. Transport of ice nuclei over the North Pacific Ocean. *Tellus 23,* 28-39.

Hobbs, P.V., H. Harrison, and E. Robinson, 1974. Atmospheric effects of pollutants. *Science 183,* 909-915.

Hoff, R.M., 1988. Vertical structure of Arctic haze observed by lidar. *J. Appl. Meteorol. 27,* 125-139.

Hoffman, D.J., and S. Solomon, 1989. Ozone destruction through heterogeneous chemistry following the eruption of El Chichon. *J. Geophys. Res. 94,* 5029-5041.

Hoppel, W.A., G.M. Frick, J.W. Fitzgerald, and R.E. Larson, 1994. Marine boundary layer measurements of new particle formation and the effects nonprecipitating clouds have on aerosol size distribution. *J. Geophys. Res. 99,* 14443-14459.

Hudson, J.G., 1989. An instantaneous CCN spectrometer. *J. Atmos. Oceanic Technol. 6,* 1055-1065.

Hudson, J.G., 1991. Observations of anthropogenic cloud condensation nuclei. *Atmos. Environ. 25A,* 2449-2455.

Hudson, J.G., and H. Li, 1995. Microphysical contrasts in Atlantic stratus. *J. Atmos. Sci. 52,* 3031-3040.

Hunter, D.E., S.E. Schwartz, R. Wagoner, and C.M. Benkovitz, 1993. Seasonal latitudinal and secular variations in temperature trend; evidence for influence of anthropogenic sulfate. *Geophys. Res. Lett. 20,* 2455-2458.

Husar, R.B., J.M. Holloway, and D.E. Patterson, 1981. Spatial and temporal pattern of eastern U.S. haziness: A summary. *Atmos. Environ. 15,* 1919-1928.

IGAC, International Global Atmospheric Chemistry Project, 1995a. Southern Hemisphere Marine Aerosol Characterization Experiment (ACE-1). Radiative Effects of Aerosols in the Remote Marine Atmosphere. Final Science and Implementation Plan. Available from T. Bates, NOAA/PMEL, 7600 Sandpoint Way NE, Seattle, WA 98115 (bates@pmel.noaa.gov).

IGAC, International Global Atmospheric Chemistry Project, 1995b. North Atlantic Regional Aerosol Characterization Experiment (ACE-2). Radiative Forcing due to Anthropogenic Aerosols over the North Atlantic Region. Science and Implementation Plan. European Commission Directorate-General XIII, Telecommunications, Information Market and Exploitation of Research, L-2920 Luxembourg. CL-NA-16229-EN-C. Luxembourg: Office for Official Publications of the European Communities.

Ignatov, A.M., L.L. Stowe, S.M. Sakerin, and G.K. Korotaev, 1995. Validation of the NOAA/ NESDIS satellite aerosol product over the North Atlantic in 1989. *J. Geophys. Res. 100*, 5123-5132.
IPCC, Intergovernmental Panel on Climate Change, 1992. The Supplementary Report to the IPCC Scientific Assessment. In *Climate Change,* Cambridge University Press, New York, 200 pp.
IPCC, Intergovernmental Panel on Climate Change, 1995a. Radiative forcing of climate change. In *Climate Change 1994*, Cambridge University Press, New York.
IPCC, Intergovernmental Panel on Climate Change, 1995b. IPCC Working Group I, *1995 Summary for Policy Makers.* IPCC WG I Technical Support Unit, Cambridge, U.K., 7 pp.
Jones, A., D.L. Roberts, and A. Slingo, 1994. A climate model study of indirect radiative forcing by anthropogenic sulphate aerosols. *Nature 370,* 450-453.
Jones, M.R., K.H. Leong, M.Q. Brewster, and B.P. Curry, 1994. Inversion of light scattering measurements for particle size and optical constants: Experimental study. *Appl. Opt. 33*, 4035-4041.
Junge, C.E., 1963. *Air Chemistry and Radioactivity.* Academic Press, New York, 382 pp.
Junge, C.E., 1975. The possible influence of aerosols on the general circulation and climate and possible approaches for modelling. Appendix 10 of *The Physical Basis of Climate and Climate Modelling*. GARP Publications, Series No. 16, World Meteorological Organization, Geneva.
Karl, T.R., R.W. Knight, G. Kukla, and J. Gavin, 1995. Evidence for radiative effects of anthropogenic sulfate aerosols in the observed climate record, in *Aerosol Forcing of Climate*, R.J. Charlson and J. Heintzenberg (eds.). John Wiley & Sons, New York.
Kaufman, Y.J., and T. Nakajima, 1993. Effect of Amazon smoke on cloud microphysics and albedo-analysis for satellite imagery. *J. Appl. Meteorol. 32*, 729-744.
Kaufman, Y.J., R.S. Fraser, and R.L. Mahoney, 1991. Fossil fuel and biomass burning effect on climate—Heating or cooling? *J. Climate 4,* 578-588.
Kent, G.S., M.P. McCormick, and S.C. Shaffner, 1991. Global optical climatology of the free tropospheric aerosol from 1.0 µm satellite occultation measurements. *J. Geophys. Res. 96*, 5249-5267.
Kiehl, J.T., and B.P. Briegleb, 1993. The relative roles of sulfate aerosols and greenhouse gases in climate forcing. *Science 260*, 311-314.
Kiehl, J.T., and H. Rodhe, 1995. Modeling geographical and seasonal forcing due to aerosols. In *Proceedings of the Dahlem Workshop on Aerosol Forcing of Climate*, R.J. Charlson and J. Heintzenberg (eds.). John Wiley & Sons, Chichester.
Kim, Y., and R.D. Cess, 1993. Effect of anthropogenic sulfate aerosols on low-level cloud albedo over oceans. *J. Geophys. Res. 98*, 14883-14885.
Kim, Y.P., J.H. Seinfeld, and P. Saxena, 1993a. Atmospheric gas-aerosol equilibrium: I. Thermodynamic model. *Aerosol Science and Technology 19*, 157-181.
Kim, Y.P., J.H. Seinfeld, and P. Saxena, 1993b. Atmospheric gas-aerosol equilibrium: II. Analysis of common approximations and activity coefficient calculation methods. *Aerosol Science and Technology 19*, 182-198.
King, M.D., F.L. Radke, and V.P. Hobbs, 1993. Optical properties of marine stratocumulus clouds modified by ships. *J. Geophys. Res. 98*, 2729-2739.
Kinne, S., O.B. Toon, and M. Prather, 1992. Buffering of stratospheric circulations by changing amounts of tropical ozone: A Pinatubo case study. *Geophys. Res. Lett. 19,* 1927-1930.
Kleinman, L.I., and P.H. Daum, 1991. Vertical distribution of aerosol particles, water vapor, and insoluble trace gases in convectively mixed air. *J. Geophys. Res. 96,* 991-1005.
Knollenberg, R.G., K. Kelly, and J.C. Wilson, 1993. Measurements of high number densities of ice crystals in the tops of tropical cumulonimbus. *J. Geophys. Res. 98*, 8639-8664.
Kondrat'yev, K. Ya., V.I. Binenko, and O.P. Petremechuk, 1981. Radiation properties of clouds subject to the anthropogenic effect of a city. *Isv. Acad. Sci. USSR Atmos. Oceanic Phys.* (Engl. Transl.) *17*, 167-174.

Kovalev, V.A., 1993. Lidar measurements of the vertical aerosol extinction profiles with range-dependent backscatter to extinction ratios. *Appl. Opt. 32*, 6053-6065.
Kumai, W., 1951. Electron-microscope study of snow-crystal nuclei. *J. Meteorol. 8*, 151-156.
Lacis, A.A., 1995. Climate forcing, climate sensitivity, and climate response. In *Proceedings of the Dahlem Workshop on Aerosol Forcing*, R.J. Charlson and J. Heintzenberg (eds.). John Wiley & Sons, Chichester.
Lacis, A.A., J.E. Hansen, and M. Sato, 1992. Climate forcing by stratospheric aerosols. *Geophys. Res. Lett. 19*, 1607-1610.
Lambert, A., R.G. Grainger, J.J. Remidos, C.D. Rodgers, M. Corney, and F.W. Taylor, 1993. Measurements of the evolution of the Mt. Pinatubo aerosol cloud by ISAMS. *Geophys. Res. Lett. 20*, 1287-1290.
Langner, J., and H. Rodhe, 1991. A global three-dimensional sulfur cycle. *J. Atmos. Chem. 13*, 225-263.
Leaitch, W.R., J.W. Strapp, G.A. Isaac, and J.G. Hudson, 1986. Cloud droplet nucleation and cloud scavenging of aerosol sulphate in polluted atmospheres. *Tellus 38B*, 328-344.
Leaitch, W.R., G.A. Isaac, J.W. Strapp, C.M. Banic, and H.A. Wiebe, 1992. The relationship between cloud droplet number concentrations and anthropogenic pollution: Observations and climatic implications. *J. Geophys. Res. 97*, 2463-2474.
Legrand, M., 1995. Atmospheric chemistry changes versus past climate inferred from polar ice cores. Pp. 123-151 in *Aerosol Forcing of Climate*, R.J. Charlson and J. Heintzenberg (eds.). John Wiley & Sons, Chichester.
Lelieveld, J., and J. Heintzenberg, 1992. Sulfate cooling effect on climate through in-cloud oxidation of anthropogenic SO_2. *Science 258*, 117-120.
Liepert, B., P. Fabian, and H. Grassl, 1994. Solar radiation in Germany—Observed trends and an assessment of their causes; Part I: Regional approach. *Beitr. Phys. Atmos. 67*, 15-29.
Lin, X., B.A. Ridley, J. Walega, G.F. Hübler, S.A. McKeen, E.-Y. Hsie, M. Trainer, F.C. Fehsenfeld, and S.C. Liu, 1994. Parameterization of subgrid scale convective cloud transport in a mesoscale regional chemistry model. *J. Geophys. Res. 99*, 25615-25630.
Liousse, C., J.E. Penner, C. Chuang, J.J. Walton, H. Eddleman, and H. Cachier, 1995. A three-dimensional model study of carbonaceous aerosols. *J. Geophys. Res.* (in press).
Malm, W.C., J.F. Sisler, D. Huffman, R.A. Eldred, and T.A. Cahill, 1994. Spatial and seasonal trends in particle concentration and optical extinction in the United States. *J. Geophys. Res. 99*, 1347-1370.
Manabe, S., and R.T Wetherald, 1980. On the distribution of climate change resulting from increase in CO_2 content of the atmosphere. *J. Atmos. Sci.37*, 99-118.
Martin, G.M., D.W. Johnson, and A. Spice, 1994. The measurement and parameterization of effective radius of droplets in warm stratiform clouds. *J. Atmos. Sci. 51*, 1823-1842.
Mather, G.K., 1991. Coalescence enhancement in large multicell storms caused by emissions from a Kraft paper mill. *J. Appl. Meteorol. 30*, 1134-1146.
McCormick, M.P., P. Hamill, T.J. Pepin, W.P. Chu, T.J. Swissler, and L.R. McMaster, 1979. Satellite studies of the stratospheric aerosol. *Bull. Am. Meteorol. Soc. 60*, 1038-1046.
McCormick, M.P., L.W. Thomason, and C.R. Trepte, 1995. Atmospheric effects of the Mount Pinatubo eruption. *Nature 373*, 399-403.
McInnes, L.M., D.S. Covert, P.K. Quinn, and M.S. Germani, 1994. Measurements of chloride depletion and sulfur enrichment in individual sea-salt particles collected from the remote marine boundary layer. *J. Geophys. Res. 99*, 8257-8268.
Meng, Z., and J.H. Seinfeld, 1994. On the source of submicrometer droplet mode of urban and regional models. *Aerosol Science and Technology 20*, 253-265.
Mészáros, E., 1992. Structure of continental clouds before the industrial era: A mystery to be solved. *Atmos. Environ. 26A*, 2469-2470.
Minnis, P., 1994. Radiative forcing by the 1991 Pinatubo eruption. Pp. J9-11 in *Preprints of Eighth Conference on Atmospheric Radiation,* American Meteorological Society, Boston.

Mitchell, J.F.B., T.C. Johns, J.M. Gregory, and S.F.B. Tett, 1995. Climate response to increasing levels of greenhouse gases and sulfate aerosols. *Nature 376*, 501-504.

Mossop, S.C., 1978. The influence of drop size distribution on the production of secondary ice particles during graupel growth. *Q. J. Roy. Meteorol. Soc. 104*, 1-13.

Müller, J.-F., and G. Brasseur, 1995. IMAGES: A three-dimensional chemical transport model of the global troposphere. *J. Geophys. Res. 100*, 16445-16490.

Nemesure, S., R. Wegener, and S.E. Schwartz, 1995. Direct shortwave forcing of climate by anthropogenic sulfate aerosol: Sensitivity to particle size, composition and relative humidity. *J. Geophys. Res.*, in press.

Neumann, H.H., and G. den Hartog, 1985. Eddy correlation measurements of atmospheric fluxes of ozone, sulphur, and particles during the Champaign intercomparison study. *J. Geophys. Res. 90*, 2097-2110.

Nicholson, K.W., 1988. Dry deposition of small particles: A review of experimental measurements. *Atmos. Environ. 22*, 2653-2666.

Norris, J.R., and C.B. Leovy, 1994. Interannual variability in stratiform cloudiness and sea surface temperature. *J. Climate 7*, 1915-1925.

Novakov, T., and C.E. Corrigan, 1995. Influence of sample composition on aerosol organic and black carbon determinations. *Proceedings of Chapman Conference on Biomass Burning*, J. Levine (ed.). MIT Press, Cambridge, Mass. (accepted).

Novakov, T., C. Rivera-Carpio, J.E. Penner, and C.F. Rogers, 1994. The effect of anthropogenic sulfate aerosols on marine cloud droplet concentrations. *Tellus 46B*, 132-141.

O'Dowd, C.D., and M.H. Smith, 1993. Physicochemical properties of aerosols over the Northeast Atlantic—Evidence for wind-speed-related submicron sea-salt aerosol production. *J. Geophys. Res. 98*, 1137-1149.

Ogren, J.A., 1995. In situ observations of aerosol properties. In *Proceedings of the Dahlem Workshop on Aerosol Forcing*, R.J. Charlson and J. Heintzenberg (eds.). John Wiley and Sons, Chichester.

Ouimette, J.R., and R.C. Flagan, 1982. Extinction coefficient of multicomponent aerosols. *Atmos. Environ. 16*, 2405-2419

Pandis, S.N, J.H. Seinfeld, and C. Pilinis, 1990. The smog-fog-smog cycle and acid deposition. *J. Geophys. Res. 95*, 18489-18500.

Pandis, S.N., A.S. Wexler, and J.H. Seinfeld, 1993. Secondary organic aerosol formation and transport: II. Predicting the ambient secondary organic aerosol size distribution. *Atmos. Environ. 27A*, 2403-2416.

Pandis, S.N., A.S. Wexler, and J.H. Seinfeld, 1995. Dynamics of tropospheric aerosols. *J. Phys. Chem. 99*, 9646-9659.

Parungo, F., E. Ackerman, W. Caldwell, and H.W. Weickmann, 1979. Individual particle analysis of Antarctic aerosols. *Tellus 31*, 521-529.

Penner, J.E., 1995. Carbonaceous aerosols influencing atmospheric radiation: Black and organic carbon. In *Proceedings of the Dahlem Workshop on Aerosol Forcing of Climate*, R.J. Charlson and J. Heintzenberg (eds.). John Wiley & Sons, Chichester.

Penner, J.E., R. Dickinson, and C. O'Neil, 1992. Effects of aerosol from biomass burning on the global radiation budget. *Science 256*, 1432-1434.

Penner, J.E., H. Eddleman, and T. Novakov, 1993. Towards the development of a global inventory of black carbon emissions. *Atmos. Environ. 27A*, 1277-1295.

Penner, J.E., C.A. Atherton, and T.E. Graedel, 1994a. Global emissions and models of photochemically active compounds. Pp. 223-248 in *Global Atmospheric-Biospheric Chemistry*, R. Prinn (ed.). Plenum Publishing, New York.

Penner, J.E., R.J. Charlson, J.M. Hales, N. Laulainen, R. Leifer, T. Novakov, J. Ogren, L.F. Radke, S.E. Schwartz, and L. Travis, 1994b. *Quantifying and Minimizing Uncertainty of Climate Forcing by Anthropogenic Aerosols*. DOE/NBB-0092T, UC-402, available from NTIS, Springfield, Va. Also *Bull. Am. Meteorol. Soc. 75*, 375-400, 1994.

Peters, K., and R. Eiden, 1992. Modeling the dry deposition velocity of aerosol particles to a spruce forest. *Atmos. Environ. 26A*, 2555-2564.

Pham, M.H., 1994. Modélisation du cycle des composés soufrés atmospheriques. Processus physico-chimiques et balans a l'échelle global. Thèse de Doctorat, l'Université de Paris, 286 pp.

Pickering, K.E., A.M. Thompson, W.-K. Tao, R.B. Rood, D.P. McNamara, and A.M. Molod, 1995. Vertical transport by convective clouds: Comparisons of three modelling approaches. *Geophys. Res. Lett. 22*, 1089-1092.

Pilinis, C., and J.H. Seinfeld, 1988. Development and evaluation of an Eulerian photochemical gas-aerosol model. *Atmos. Environ. 22*, 1985-2001.

Pilinis, C., S.N. Pandis, and J.H. Seinfeld, 1995. Sensitivity of direct climate forcing by atmospheric aerosols to aerosol size and composition. *J. Geophys. Res.*, in press.

Pincus, R., and M.B. Baker, 1994. Effect of precipitation on the albedo susceptibility of clouds in the marine boundary layer. *Nature 372*, 250-252.

Pollack, J.B., and T.P. Ackerman, 1983. Possible effects of the El Chichon cloud on the radiation budget of the northern tropics. *Geophys. Res. Lett. 10*, 1057-1060.

Pruppacher, H.R., 1995. A new look at ice nucleation in supercooled water drops. *J. Atmos. Sci. 52*, 1924-1933.

Pruppacher, H.R., and J.D. Klett, 1978. Pp. 229-230 in *Microphysics of Clouds and Precipitation*. D. Reidel, Dordrecht, Holland, 714 pp.

Pueschel, R.F., C.C. Van Valin, R.G. Castillo, J.A. Kadlecek, E. Ganor, 1986. Aerosols in polluted versus nonpolluted air masses: Long-range transport and effects on clouds. *J. Climate Appl. Meteorol. 25*, 1908-1917.

Quinn, P.K., D.S. Covert, T.S. Bates, V.N. Kapustin, D.C. Ramsey-Bell, and L.M. McInnes, 1993. Dimethylsulfide/cloud condensation nuclei/climate system: Relevant size-resolved measurements of the chemical and physical properties of the atmospheric aerosol particles. *J. Geophys. Res. 98*, 10411-10427.

Radke, L.F., and P.V. Hobbs, 1976. Cloud condensation nuclei on the Atlantic seaboard of the United States. *Science 193*, 999-1002.

Radke, L.F., J.A. Coakley, Jr., and M.D. King, 1989a. Direct and remote sensing observations of the effects of ships on clouds, *Science 246*, 1146-1149.

Radke, L.G., C.A. Brock, J.H. Lyons, P.V. Hobbs, and R.C. Schnell, 1989b. Aerosol and lidar measurements of hazes in mid-latitude and polar airmasses. *Atmos. Environ. 23*, 2417-2430.

Radke, L.F., D.A. Hegg, P.V. Hobbs, J.D. Nance, J.H. Lyons, K.K. Laursen, R.E. Weiss, P.J. Riggan, and D.E. Ward, 1991. Particulate and trace gas emissions from large biomass fires in North America. Pp. 209-228 in *Global Biomass Burning*, J.S. Levine (ed.). MIT Press, Cambridge, Mass., 569 pp.

Radke, L.F., D.A. Hegg, J.H. Lyons, and J.D. Nance, 1992. Scavenging of smokes by biomass fire-capping cumulus clouds. Pp. 371-380 in *Proceedings of the Fifth International Conference on Precipitation Scavenging and Atmospheric-Surface Exchange Processes*, S.E. Schwartz and W.G.N. Slinn (eds.).

Raes, F., and R. Van Dingenen, 1995. Comment on "The relationship between DMS flux and CCN concentration in remote marine regions" by S.N. Pandis, L.M. Russell, and J.H. Seinfeld. *J. Geophys. Res. 100*, 14355-14356.

Raga, G.B., and P.R. Jonas, 1993. On the link between cloud-top radiative properties and subcloud aerosol concentrations. *Q. J. Roy. Meteorol. Soc. 119*, 1419-1425.

Ramaswamy, V., R.J. Charlson, J.A. Coakley, J.L. Gras, Harshvardhan, G. Kukla, M.P. McCormick, D. Moller, E. Roeckner, L.L. Stowe, and J. Taylor, 1995. What are the observed and anticipated meteorological and climatic responses to aerosol forcing? Pp. 384-399 in *Aerosol Forcing of Climate*, R.J. Charlson and J. Heintzenberg (eds.). Wiley and Sons, Chichester, U.K.

Rao, C.R.N., L. Stowe, and E.P. McClain, 1989. Remote sensing of aerosols over the oceans using AVHRR data: Theory, application and applications. *Intl. J. Remote Sensing 4*, 743-749.
Robock, A., and J. Mao, 1995. The volcanic signal in surface temperature observations. *J. Climate*, in press.
Roche, A.E., J.B. Kumer, J.L. Mergenthaler, G.A. Ely, W.G. Uplinger, J.F. Potter, T.C. James, and L.W. Sterritt, 1993. The Cryogenic Limb Array Etalon Spectrometer (CLAES) on UARS: Experiment description and performance. *J. Geophys. Res. 98*, 10763.
Roeckner, E., T. Siebert, and J. Feichter, 1995. Climatic response to anthropogenic sulfate forcing simulated with a general circulation model. In *Proceedings of the Dahlem Workshop on Aerosol Forcing of Climate*, R.J. Charlson and J. Heintzenberg (eds.). John Wiley & Sons, Chichester.
Rojas, C.M., R.E. Van Grieken, and R.W. Laane, 1993. Comparison of three dry deposition models applied to field measurements in the Southern Bight of the North Sea. *Atmos. Environ. 27A*, 363-370.
Russell, L.M., S.N. Pandis, and J.H. Seinfeld, 1994. Aerosol production and growth in the marine boundary layer. *J. Geophys. Res. 99*, 20989-21003.
Russell, P.B., and M.P. McCormick, 1989. SAGE II aerosol data validation and initial data use: An introduction and overview. *J. Geophys. Res. 94*, 8353-8366.
Santer, B.D., K.E. Taylor, T.M.L. Wigley, J.E. Penner, U. Cubasch, and P.D. Jones, 1995a. Towards the detection and attribution of anthropogenic effect on climate. *Climate Dynamics*, in press.
Santer, B.D., K.E. Taylor, T.M.L. Wigley, P.D. Jones, D.J. Karoly, J.F.B. Mitchell, A.H. Oort, J.E. Penner, V. Ramaswamy, M.D. Schwarzkopf, R.J. Stouffer, and S.F.B. Tett, 1995b. A search for human influence on the thermal structure of the atmosphere. *Nature* (submitted August 1995).
Sassen, K., 1992. Evidence for liquid-phase cirrus cloud formation from volcanic aerosols: Climatic implications, *Science 257*, 516-519.
Schnell, R.C., and G. Vali, 1972. Atmospheric ice nuclei from decomposing vegetation. *Nature 236*, 163-165.
Schnell, R.C., and G. Vali, 1976. Biogenic ice nuclei; Part I. Terrestrial and marine sources. *J. Atmos. Sci. 33*, 1554-1564.
Schwartz, S.E., 1988. Are the global cloud albedo and climate controlled by marine phytoplankton? *Nature 336*, 441-445.
Schwartz, S.E., and W.G.N. Slinn, 1992. *Precipitation Scavenging and Atmosphere-Surface Exchange*, Vol. 1—*The Georgii Volume: Precipitation Scavenging Processes*; Vol. 2—*The Semonin Volume: Atmosphere-Surface Exchange Processes*; Vol. 3—*The Summers Volume: Applications and Appraisals*, Hemisphere Publishing Co., Washington, D.C., 1808 pp.
Scorer, R.S., 1987. Ship trails. *Atmos. Environ. 21*, 1417-1425.
Slinn, W.G.N., L.F. Radke, and P.C. Katen, 1983. Inland transport, mixing, and dry deposition of sea-salt particles. Pp. 1037-1046 in *Precipitation Scavenging, Dry Deposition, and Resuspension*, Vol. 2, H.R. Pruppacher, R.G. Semonin, and W.G.N. Slinn (coordinators). Elsevier, New York.
Solomon, S., R.W. Portman, R.R. Garcia, L.W. Thomason, L.R. Poole, and M.P. McCormick, 1995. The role of aerosol variability in the anthropogenic ozone depletion at northern mid-latitudes. *J. Geophys. Res.*, in press.
Stephens, G.L., 1987. On the effects of ice crystal porocity on the characteristics of cirrus clouds. *J. Geophys. Res. 92*, 3979-3984.
Stephens, G.L., S.C. Tsay, P.W. Stockhouse, and P.J. Flatau, 1990. The relevance of the microphysical and radiative properties of cirrus clouds to climate and climate feedback. *J. Atmos. Sci. 47*, 1742-1753.

Tang, I.N., and R.H. Munkelwitz, 1993. Composition and temperature dependence of the deliquescence properties of hygroscopic aerosols. *Atmos. Environ.* 27A, 467-473.

Tang, I.N., and R.H. Munkelwitz, 1994a. Water activities, densities, and refractive indices of aqueous sulfates and sodium nitrate droplets of atmospheric importance. *J. Geophys. Res.* 99, 18801, 18808.

Tang, I.N., and R.H. Munkelwitz, 1994b. Aerosol phase transformation and growth in the atmosphere. *J. Appl. Meteorol.* 33, 791-796.

Taylor, K.E., and J. Penner, 1994. Response of the climate system to atmospheric aerosols and greenhouse gases. *Nature 369*, 734-737.

Tegen, I., and I. Fung, 1995. Modeling of mineral dust in the atmosphere: Sources, transport and optical thickness. *J. Geophys. Res.*, in press.

Thomason, L.W., and L.R. Poole, 1993. Use of atmospheric aerosol properties as diagnostics of Antarctic vortex processes. *J. Geophys. Res.* 98, 23003-23012.

Thompson, A.M., K.E. Pickering, R.R. Dickerson, W.G. Ellis, Jr., D.J. Jacob, J.R. Scala, W.-K. Tao, D.P. McNamara, and J. Simpson, 1994. Convective transport over the central United States and its role in regional CO and ozone budgets. *J. Geophys. Res. 99*, 18703-18711.

Twomey, S., 1971. Figure 8.9 in *The Report of the Study of Man's Impact on Climate (SMIC). Inadvertent Climate Modification.* MIT Press, Cambridge, Mass.

Twomey, S., 1977. *Atmospheric Aerosols.* Elsevier, 302 pp.

Twomey, S., 1991. Aerosols, clouds, and radiation. *Atmos. Environ. 25A*, 2435-2442.

Twomey, S., A.K. Davidson, and K.J. Seton, 1978. Results of five years' observations of cloud nucleus concentration at Robertson, New South Wales. *J. Atmos. Sci. 35*, 650-656.

Vali, G., 1975. Summary, Ice Nucleation Workshop, 1975, Laramie, Wyoming. *Bull. Am. Meteorol. Soc. 56*, 1180-1181.

Vali, G., 1985. Atmospheric ice nucleation—A review. *J. Rech. Atmos. 19*, 105-115.

Waggoner, A.P., R.E. Weiss, N.C. Ahlquist, D.S. Covert, S. Will, and R.J. Charlson, 1981. Optical characteristics of atmospheric aerosols. *Atmos. Environ. 15*, 1891-1909.

Wallace, J.M., and P.V. Hobbs, 1977. *Atmospheric Science, An Introductory Survey.* Academic Press, New York, 467 pp.

Wallace, J.M., Y. Zhang, and I. Bajuk, 1995. Interpretation of interdecadal trends in Northern Hemispheric surface air temperature. *J. Climate*, in review.

Warner, J., and S. Twomey, 1967. The production of cloud nuclei by cane fires and the effect on cloud droplet concentration. *J. Atmos. Sci. 24*, 704-706.

Warren, D.R., and J.H. Seinfeld, 1985. Simulation of aerosol size distribution evolution in systems with simultaneous nucleation, condensation, and coagulation. *Aerosol Science and Technology 4*, 31-43.

WCRP, World Climate Research Programme, 1991. *Radiation and Climate: Second Workshop on Implementation of the Baseline Surface Radiation Network.* Report WCRP-64, WMO/TD-No. 453. World Meteorological Organization, Geneva.

Weber, R.J., P.H. McMurry, F.L. Eisele, and D.J. Tanner, 1995a. Measurement of expected nucleation precursor species and 5 to 500 nm diameter particles at Mauna Loa Observatory, Hawaii. *J. Atmos. Sci. 52*, 2242-2257.

Weber, R.J., J.J. Marti, P.H. McMurry, F.L. Eisele, D.J. Tanner, and A. Jefferson, 1995b. Measured atmospheric new particle formation rates: Implications for nucleation mechanisms. *Chem. Engg. Comm.* (in press).

Wexler, A.S., and J.H. Seinfeld, 1991. Second-generation inorganic aerosol model. *Atmos. Environ. 25A*, 2731-2748.

Wexler, A.S., F.W. Lurmann, and J.H. Seinfeld, 1994. Modeling urban and regional aerosols: I. Model development. *Atmos. Environ. 28*, 531-546.

White, W., 1990. *Contributions to Light Extinction*, Section 4. Report 24, U.S. National Acid Precipitation Assessment Program. U.S. Government Printing Office, Washington, D.C.

Wigley, T.M.L., 1989. Possible climate change due to SO_2 derived cloud condensation nuclei. *Nature 339*, 365-367.
Wiscombe, W.J., and G.W. Grams, 1976. The backscattered fraction in two-stream approximations. *J. Atmos. Sci. 33*, 2440-2451.
WMO, World Meteorological Organization, 1989. *Report of the International Ozone Trends Panel, Global Ozone Research and Monitoring Project No. 18.* World Meteorological Organization, Geneva.
WMO, World Meteorological Organization, 1992. *Scientific Assessment of Ozone Depletion; 1991, Global Ozone Research and Monitoring Project No. 25.* World Meteorological Organization, Geneva.

A

Illustrations of Recommended Research, Emphasizing Recent Literature

SMIC, Study of Man's Impact on Climate (1971). *Inadvertent Climate Modification.* Report of the Study of Man's Impact on Climate. MIT Press, Cambridge, Mass., 308 pp.:

> We recommend the compilation of figures on global particle production rates that are accurate to a factor of 2 or better . . . . All studies should give information on mass fluxes as well as size distribution in the range between 0.01 and 10 μm radius.
>
> We recommend that the transformations of atmospheric trace gases which lead to particle formation be studied. [These studies] should include the collection of better data on the distribution of these gases in the atmosphere and on their life cycles and residence times.
>
> We recommend periodic measurement (for example, at intervals of 2 years) of the major sources of particles which are man-made and over which he can exert control.
>
> We recommend more comprehensive studies of the relative importance of the principal removal mechanisms of particles and gases by precipitation, of particles by impaction, sedimentation, and diffusion at the ground, and of gases by absorption at the ground. This has a particularly important bearing on computation of residences times in the atmosphere.
>
> We recommend that suitable methods be developed to measure the particle size distribution below 0.1 μm radius and to study its modification in clean and polluted atmospheres.

We recommend the spatial distributions of particle concentrations and trends with time be monitored on a global basis. A network of about 100 stations is required to give representative data for the whole atmosphere. For all optical measurements an accuracy of 5 percent is required to obtain meaningful data.

Because of the wide range of particle sizes, different methods have to be used simultaneously, in particular:

a. . . . [transmissivity]
b. . . . [horizontal extinction]
c. . . . [trends for particles < 0.1 μm]
d. . . . [oceans: continue electrical conductivity measurements]
e. Trends in the concentration of ice and cloud nuclei should be continuously monitored. Very careful selection and perhaps improvement of existing methods are necessary.

We recommend increased research on the refractive index (including absorption) of atmospheric particles in relation to their composition, origin, size, and shape and its change with increasing pollution for short- and long-wave radiation. Also, more data on particle growth with humidity are needed to understand the influence on the refractive index with relative humidity.

We recommend comprehensive comparative studies of the radiation fields for clean and polluted atmospheres in order better to identify the effects of short- and long-wave radiation on the atmosphere and its modification with increasing pollution.

We recommend measurements of albedo and other radiative properties of clouds and fogs in unpolluted and polluted areas, in conjunction with sufficiently complete measurements of cloud microstructure. These very important measurements require operation of instrumented aircraft.

We recommend studies of the effects of pollution on the refractive index of cloud droplets and ice crystals.

We recommend theoretical study of the integrated effects of pollution on radiative properties of clouds and fogs.

We recommend that comprehensive field studies directed toward the resolution of the question of the effect of particle concentration on frequency, type, and intensity of clouds and precipitation be designed and implemented.

We recommend that at suitable time intervals (about 2 years) the refractive index and the detailed size distribution of particles be determined at selected places in both clean and polluted air.

We recommend that objective methods for monitoring cloud cover by satellites or other means and for monitoring changes of cloud cover over large areas of the atmosphere be developed and implemented.

> We recommend that at suitable intervals (for example, 5 years) the optical properties of clouds (reflectivity, transmissivity, emissivity, and absorption) be determined in areas of increasing air pollution.
>
> We recommend monitoring of [sulfur and nitrogen gases] . . . in unpolluted areas, with an accuracy of ±10 percent. Discontinuous measurement at about 10 baseline stations and 100 regional stations is required in order to improve the understanding of the life history of these gases and of the particle formation processes in which they are involved.
>
> We recommend that high priority be given to efforts to determine the humidity at which cirrus cloud forms. In particular, the question of whether cirrus forms through sublimation or by means of the liquid phase should be answered.
>
> We recommend that information [be] collected on the fundamental physical properties of cirrus clouds. These include water content, particle concentration, and distribution of sizes and shapes of crystals.
>
> We recommend that the optical properties of cirrus clouds, in both solar and infrared radiation, be investigated.
>
> We recommend that high priority be given to monitoring trends, if any, in cirrus cloudiness and characteristics. For this purpose, objective methods are needed in order to distinguish between cirrus and lower clouds.

Hobbs, P.V., H. Harrison, and E. Robinson (1974). Atmospheric effects of pollutants. *Science 183,* 909-915:

> It is our opinion . . . that coupled effects between particles and clouds are likely to outweigh direct albedo effects from the particles themselves.
>
> Most critical of all, do the pollutants affect aerosols which play a role in cloud processes? If so, the structure and distribution of clouds may be affected by the pollutants, thereby causing changes in precipitation and optical scattering. Research priority should be given to this area.
>
> Junge, C.E. (1975). The possible influence of aerosols on the general circulation and climate and possible approaches for modeling. Chapter 10 of *The Physical Basis of Climate and Climate Modeling.* GARP Publication Series No. 16, World Meteorological Organization, Geneva:
>
> Understanding of the indirect influence [of aerosols] on the short- and long-wave radiation budget by modification of cloud micro- and macrostructure, i.e., albedo, cloud cover, etc. Problem practically open.
>
> The available data indicate that the indirect effects of aerosols on water clouds may be very important for changes in the radiation budget, perhaps more than the direct effects of aerosols. Unfortunately, no attempts to estimate these effects have been made so far.

Shaw, G.E. (1987). Aerosol as climate regulators: A climate-biosphere linkage? *Atmos. Env. 21*, 985-986:

> Of the particles in the atmosphere, those around a few tenths of a μm are noteworthy. Firstly, they are relatively immune to removal and therefore remain suspended for long periods, and secondly, they interact strongly with sunlight. The collective system of sub-μm aerosols, therefore, constitutes an enormously sensitive climate regulating machine. In this regard, it is interesting to note that the present quantity of biologically produced sulfate aerosol is nearly that needed to opalize the atmosphere. This might suggest that the sulfur aerosol system plays a role in climate . . . . We should try and find evidence pro or con regarding the past record of atmospheric transparency to sunlight.

Charlson, R.J., J.E. Lovelock, M.O. Andreae, and S.G. Warren (1987). Oceanic phytoplankton, atmospheric sulfur, cloud albedo, and climate. *Nature 326*, 655-661:

> There are significant gaps in our knowledge of this proposed feedback system. Most importantly, we need to understand the climatic factors affecting DMS [dimethyl sulfide] emission. Because some species produce much more DMS than others, we must include the necessary understanding of controls on phytoplankton species abundance. We also need to understand the relationship between DMS concentration in the air and the CCN population, through the intervening aerosol physical processes. Knowing how the area of cloud cover is influenced by CCN [cloud condensation nuclei] is also important.

Charlson, R.J. (1988). Have the concentrations of tropospheric aerosol particles changed? Pp. 79-90 in *The Changing Atmosphere*, F.S. Rowland and I.S.A. Isaksen (eds.). Wiley, New York:

> In order to understand and be able to predict the aerosol particle concentrations and effects of the future, especially with regard to their climatic role as CCN, we will require more and substantially different measurements than are presently being made. Full aerosol characterizations (including chemical, physical, optical, and cloud nucleating characteristics) should be conducted systematically in areas that are frequently downwind of industrial and other key source regions. Chemical data are required for understanding chemical effects and for relating to sources. Physical data are needed both to understand the atmospheric processes and turnover times of the aerosol and to understand physical effects. These measurements should be undertaken with the intent of establishing the extent of anthropogenic increases of aerosol concentration, to document trends, and eventually to be able to forecast effects.

Bigg, E.K. (1986). Discrepancy between observation and prediction of concentrations of cloud condensation nuclei. *Atmos. Res.* 20, 82-86:

> Obviously we need to know how reliable these measurements are and how well they relate to droplet growth in the free atmosphere before we can have full confidence in the calculations.

Gras, J.L. (1990). Cloud condensation nuclei over the southern ocean. *Geophys. Res. Lett.* 17, 1565-1567:

> [T]o answer questions regarding possible long-term climate effects of CCN concentration changes, either through anthropogenic activities or from marine DMS emissions, far better climatologies than are currently available are required from a range of remote oceanic regions. Thorough intercomparison is clearly an essential feature for these programs.

Grassl, H. (1988). What are the radiative and climatic consequences of the change in concentration of atmospheric aerosol particles? Pp. 187-199 in *The Changing Atmosphere*, F.S. Rowland and I.S.A. Isaksen (eds.). Wiley, New York:

> The strong dependence of the local planetary albedo change on aerosol particle size distribution change and soot content underlines the need for more reliable input parameters and more sophisticated models before the sign and the relative magnitude of an aerosol climate signal . . . may be given with higher reliability . . . . Further steps in aerosol research related to climate should be simultaneous measurements of aerosol particle parameters and optical cloud parameters, as well as the introduction of aerosol transport into atmospheric general circulation models.

Crutzen, P.J., and M.O. Andreae (1990). Biomass burning in the tropics: Impact on atmospheric chemistry and biogeochemical cycles. *Science 250*, 1669-1678:

> Because of the great importance of biomass burning and deforestation activities for climate, atmospheric chemistry, and ecology, it is clearly of the utmost importance to improve considerably our quantitative knowledge of these processes . . . . Biomass burning is also an important source of smoke particles, a large amount (maybe all) of which act as CCN or can be converted to CCN by atmospheric deposition of hygroscopic substances. The amount of aerosols produced from biomass burning is comparable to that of anthropogenic sulfate aerosol. Through this process, the cloud microphysical and radiative processes in tropical rain and cloud systems can be affected with potential climatic and hydrological consequences.

Kaufman, Y.J., and T. Nakajima (1993). Effect of Amazon smoke on cloud microphysics and albedo—Analysis from satellite imagery. *J. Appl. Meteorol.* 32, 729-744:

> [T]he presence of dense smoke (an increase in the optical thickness from 0.1 to 2.1) can reduce the remotely sensed drop size of continental cloud

drops from 15 to 9 μm. Due to both the high initial reflectance of clouds in the visible part of the spectrum and the presence of graphitic carbon, the average cloud reflectance at 0.64 μm is reduced from 0.71 to 0.68 . . . . [H]igh concentration of aerosol particles [from biomass burning] causes a decrease in the cloud-drop size and . . . darkens the bright Amazonian clouds . . . . [I]t is possible to explain the reduction in the cloud reflectance . . . for smoke imag[inary index of refraction] of -0.02 to -0.03.

The increase in the average cloud-top temperature as a function of the smoke optical thickness indicates the possibility that reduction of convection due to the smoke absorption and reflection of sunlight caused a decrease in the updraft speed and in the amount of liquid water available to form the cloud . . . . [T]he results indicate that smoke reduces rather than increases the reflectivity of clouds in the topics, in contrast to previous assumptions . . . . [T]o better understand the interaction of aerosol particles with clouds, more information about the details of their size distribution and time evolution are required.

Novakov, T., and J.E. Penner (1993). Large contribution of organic aerosols to CCN concentrations. *Nature 365,* 823-826:

The apparent ability of organic aerosols to serve as CCN can have two possible . . . explanations . . . . [H_2SO_4 condensed on organics, or the organics are CCN]. The present data, however, are insufficient to argue for either of these possibilities.

Hansen, J.E., and A.A. Lacis (1990). Sun and dust versus greenhouse gases: An assessment of their relative roles in global climate change. *Nature 346,* 713-719:

We conclude that the lack of global aerosol data makes it impossible at present to determine the net anthropogenic aerosol forcing of the climate system . . . . Satisfactory quantitative analysis of the net climate forcing owing to anthropogenic aerosols will be difficult, because of the inhomogeneous distribution of the aerosols. It will be necessary to monitor global tropospheric aerosol properties, and carry out in situ case studies under a broad variety of conditions. Such data could yield the direct aerosol climate forcing, and in conjunction with global cloud data, it could also allow evaluation of aerosol-cloud interactions.

Aerosols are the source of our greatest uncertainty about climate forcing . . . . Tropospheric aerosols are difficult to monitor because of their spatial inhomogeneity, but they are a crucial variable because of the strong anthropogenic influence on their amount. Not only will it be necessary to monitor the aerosols but also to have continued global cloud observations, because of possible interactions between aerosols and clouds . . . . [U]ntil the research on aerosols is carried out (a difficult task), we do not even know the direction of the change of climate forcing on decadal time scales which would be caused by a modification of fossil-fuel use.

Langner, J., and H. Rodhe (1991). A global three-dimensional model of the tropospheric sulfur cycle. *J. Atmos. Chem. 13,* 225-263:

> Many assumptions have been made in deriving the model and further work is needed to narrow the uncertainties in model parameters. Further improvements in our understanding of the circulation of sulfur species through the global atmosphere require that the uncertainties be narrowed of, in particular, the following fluxes/factors:
>
> - the emission of DMS from the oceans,
> - the oxidation pathways of DMS,
> - the liquid phase oxidation of SO_2, and
> - the wet removal of aerosol sulfate.
>
> [They ignored cloud venting of sulfate and their dry deposition velocity for sulfate has essentially zero experimental support.]
>
> Possible approaches to this challenge include field campaigns in marine environments under well-defined meteorological conditions with simultaneous measurements of the various sulfur species and H_2O_2 at different heights . . . supported by models that describe boundary-layer mixing, surface exchange, cloud processes, and chemistry.

Savoie, D.L., J.M. Prospero, S.J. Oltmans, W.C. Graustein, K.K. Turekian, J.T. Merrill, and H. Levy II (1992). Sources of nitrate and ozone in the marine boundary layer of the tropical North Atlantic. *J. Geophys. Res. 97,* 11575-11589:

> A recent synthesis of field measurements and modeling results . . . indicates that the nitrogen and sulfur transport simulated by the models is far weaker over the open ocean than that indicated by our measurements or those of other investigators . . . . Such models are likely to play a major role in predicting how the world's weather and climate will change as a consequence of man's activities. However, to serve that purpose they must first be able to correctly simulate the transport from the major source regions as well as the resulting concentration fields for current conditions.

Langner, J., H. Rodhe, P.J. Crutzen, and P. Zimmermann (1992). Anthropogenic influence on the distribution of tropospheric sulfate aerosol. *Nature 359,* 712-716:

> Little information is available to verify the calculated changes in the tropospheric distribution of sulfate and fluxes of n.s.s. [non sea-salt] sulfate presented here. This is especially true regarding concentrations of non-sea salt sulfate in air . . . . The best quantitative information comes from ice-core records obtained in Greenland showing an enhancement by a factor of 2-4 over the last hundred years . . . .

Charlson, R.J., J. Langner, H. Rodhe, C.B. Leovy, and S.G. Warren (1991). Perturbation of the Northern Hemisphere radiative balance by backscattering from anthropogenic sulfate aerosols. *Tellus 43AB*, 152-163:

> Large remaining questions and problems:
>
> The first main category concerns the refinement of the model. Better knowledge of [the mass scattering coefficient] and [the hemispheric backscattered fraction], their variability and dependence on controlling factors, would increase confidence in the calculated effects . . . [model improvements: solar zenith angle, time of day, surface albedo, clear sky transmission]. Seasonal variation of the [sulfate burden] should be added . . . . Finally, the calculated radiative effect should be compared to measurements, both ground and satellite based. Improvements in the uncertainty of natural and anthropogenic emission fluxes of gaseous sulfur compounds are also needed, as are improvements in modeling the transformation and removal processes.
>
> We cannot be as quantitative in comparing the direct effect with the possible indirect CCN effect of anthropogenic sulfate... A mere 20% increase of CCN in the NH [Northern Hemisphere] is calculated to yield a cooling of ca. 1 W m^{-2} (Wigley, 1989). While there certainly is potential for the anthropogenic sulfate to have increased the CCN number concentration, there are as yet no data and there is no agreed-upon theory relating number concentration of CCN to mass concentrations of sulfate.

Lelieveld, J., and J. Heintzenberg (1992). Sulfate cooling effect on climate through in-cloud oxidation of anthropogenic SO_2. *Science 258*, 117-120:

> Earlier estimates of the sulfate climate forcing were based on a limited number of sulfate-scattering correlation measurements from which a high sulfate-scattering efficiency was derived. Model results suggest that cloud processing of air is the underlying mechanism. Aqueous phase oxidation of SO_2 into sulfate and the subsequent release of the dry aerosol by cloud evaporation render sulfate a much more efficient scatterer than through gas-phase SO_2 oxidation.
>
> On the basis of aircraft measurements over southern Sweden . . . , Charlson et al. . . . adopted an empirical value of alpha = 8.5 m^2 g^{-1}. This mass-scattering coefficient, derived from a correlation between sulfate amount and light-scattering, implies that sulfate is very efficient in scattering solar radiation. In these measurements, however, 40-60% of the light-scattering material measured was not sulfate. Hence, a causal relation between aerosol chemical composition and scattering properties was not demonstrated.
>
> [W]e conclude that the mean climate forcing by sulfate in this part of the globe [NH: 0.5-2.0 μg m^{-3} sulfate] may be about -0.5 to -1.0 W m^{-2}, to a large extent caused by in-cloud oxidation of anthropogenic SO_2 [versus the -1.1 W m^{-2} of Charlson et al., 1991].

Kiehl, J.T., and B.P. Briegleb (1993). The relative roles of sulfate aerosols and greenhouse gases in climate forcing. *Science 260*, 311-314:

> To define better the direct forcing due to sulfate aerosols, more comprehensive and simultaneous observational data are needed on the chemical, physical, and radiative properties of the aerosol. These data are needed for a range of different geographic locations, because the sulfate characteristics are no doubt linked to the chemical environment (e.g., NH_3 sources).
>
> [R]ecent estimates of the climate forcing due to smoke . . . use the same simple radiative model of Charlson et al., where the aerosol specific extinction is assumed to be independent of wavelength. Most aerosols exhibit a decrease in extinction with wavelength. Thus, the radiative effects of smoke have probably been overestimated . . . . The remaining aerosols of importance are those composed of elementary carbon. Estimates of the spatial effects of these aerosols on the climate system are urgently needed.

Box, M.A., and T. Trautman (1994). Computation of anthropogenic sulfate aerosol forcing using radiative perturbation theory. *Tellus 46B*, 33-39:

> It has been argued by many authors over the years that anthropogenic aerosols are likely to have a cooling effect on the earth's surface temperature, by reflecting sunlight back to space . . . . [T]o quantify that forcing, we need to complete a three-stage calculation: . . . (estimate) the mass loading of anthropogenic aerosols . . . ; this must be converted to an optical model; and then the radiative effects of this optical model must be evaluated. In this paper we have concentrated on the second and third of these stages, to determine the forcing produced per unit of model aerosol.
>
> We have found that the perturbation factor due to an additional aerosol loading with dry sulfate aerosols, as estimated by Charlson et al. (1991) is too large, by a factor of between 2 and 3. The main reason for this is that the spectral dependence of the aerosol optical properties is ignored in the method used by Charlson et al. (1991). However, when the effects of humidity were taken into account, our final results are very close to those of Charlson et al. . . . .
>
> This is in sharp contrast to the very recent calculations of Kiehl and Briegleb (1993), despite their use of a fully realistic sulfate optical model, which appears quite similar to ours. However, one key difference is in their handling of the effects of humidity, where they use a single factor to rescale the specific extinction. Our calculations . . . show that this rescaling is itself wavelength dependent . . . . Although we believe that our results are the most accurate to date, it is clear that subtle details of the sulfate aerosol model, including the effects of humidity, can have a significant effect on the final result.

Twomey, S. (1991). Aerosols, clouds, and radiation. *Atmos. Env.* 25A, 2435-2442:

> It is regrettable that present-day monitoring programs include aerosols only to the extent of the "Aitken count"; measurements of cloud-nucleating particles are not especially difficult. At the present time, little is known about the major gas-to-particle route(s) for sulfur, etc., in the cleaner parts of the atmosphere; far more anthropogenic sulfur is being cycled annually through the atmosphere than is converted from gas into new small particles, and this anthropogenic sulfur is further augmented by the biogenic sulfur injections discussed by Charlson et al. (1987) and others.
>
> Particle production does not appear to be sulfur-limited, and one cannot rule out the possibility that some other trace gas, or photon supply, might be the limiting factor. Clearly many more field measurements and laboratory experiments are called for, rather than endless repetitions of computer simulations that are closely related to each other and parameterize in very similar ways. Satellite measurements capable of giving not just cloud cover (and cloud-top temperature) but also some information about cloud microphysics would be valuable, especially if the measurements continued unaltered over long periods of time.
>
> Parenthetically, it should be pointed out that Schwartz (1988) argued that since mean albedos seemed to be about equal for Northern and Southern Hemispheres (despite there being perhaps 3-10 times more sulfur injected into the atmosphere of the Northern Hemisphere, albedo modifications by man-made emissions could be discounted. [The figure] shows [however] that it is the clean regions that are most susceptible to albedo increase: Schwartz['s] estimated 6 megatons yearly of anthropogenic S could easily have more impact in the clean regions of the SH [Southern Hemisphere] than 150 megatons in the NH.

Schneider, S.H. (1994). Detecting climatic change signals: Are there any "fingerprints"? *Science 263,* 341-347:

> As for the aerosol forcing, nobody has tried to produce a regional map of CCN-induced (or soot-induced) cloud albedo changes . . . .

Kaufman, Y.J., R.S. Fraser, and R.L. Mahoney (1991). Fossil fuel and biomass burning effect on climate—Heating or cooling? *J. Climate 4,* 578-588:

> [T]o decrease the uncertainty as to the effect of coal and oil burning on climate, there is a need to verify experimentally the relation between the presence of pollution and the corresponding change in cloud characteristics. Since the effect of pollution on clouds is usually much smaller than the variability in the cloud characteristics due to dynamic effects, the relation must be based on statistical studies of cloud characteristics and aerosol density. Satellite imagery can be used to study thousands of clouds simultaneously . . . , parallel to studies of the surrounding aerosol . . . .

Because of the nonlinearity of the relation between cloud albedo and aerosol density, in addition to detailed studies of cloud microphysics from aircraft or balloons, and its relation to aerosol concentration, there is need to study the relation between the aerosol concentration and the average cloud albedo from satellite imagery.

Leaitch, W.R., G.A. Isaac, J.W. Strapp, C.M. Banic, and H.A. Wiebe (1992). The relationship between cloud droplet number concentrations and anthropogenic pollution: Observations and climatic implications. *J. Geophys. Res. 97*, 2463-2474:

> Much of the current discussion on this issue is quite speculative due to a lack of observational data describing how the CDNC [cloud drop number concentrations] are affected by changing pollution.

> These observations offer support for the importance of the issue of pollution, cloud microphysics, and cloud albedo. They should underscore the need for more global measurements of cloud microphysics and chemistry, particularly in remote regions, to help address the issue of global climate change.

Kaufman, Y.J., and M.-D. Chou (1993). Model simulations of the competing climatic effects of SO_2 and CO_2. *J. Climate 6*, 1241-1252:

> [Not addressed: Induced changes in cloud lifetime]

> The major uncertainties concern the relationship between the SO_2 emission and the CN [condensation nuclei] production, the relationship between the CCN concentration and the cloud optical thickness, the value of the background CCN concentration, and the vertical distribution of the SO_2 derived CCN. In order to fully assess the impact of anthropogenic SO_2 on climate, we need to improve our understanding of these processes.

Leaitch, W.R., and G.A. Isaac (1994). On the relationship between sulfate and cloud droplet number concentration. *J. Climate 7*, 206-212:

> Scatter in the data makes it impossible to constrain model parameters; however, the comparisons suggest that there may not be a universal relationship, and that the uncertainties involved in trying to model this process are large.

> It is important to emphasize that the sensitivity of N_d [the number density of cloud droplets] to sulfate may be greater at lower sulfate concentrations . . . . Although our understanding of the effect of intense pollution on cloud microphysics is by no means satisfactory, efforts should be focused on obtaining more data concerning the effect of anthropogenic sulfate, and other aerosol species, at concentrations more typical of global sulfate concentrations [i.e., <50 nEq m^{-3}].

> There is a clear need to study the relationships among N_d, CCN, and sulfate in detail before parameterizations of this nature can be properly

described. To help reduce the data scatter of the type discussed here, it will be necessary to accumulate case studies of similar cloud types, but with varying sulfate contents, so that further differentiation by cloud type is possible.

The global simulations of Langner et al. (1992) indicate that only 6% of the anthropogenic sulfur is available for new aerosol particle formation, from which they questioned the assumption of Kaufman et al. (1991) that 40% of the fossil fuel SO_2 emissions resulted in new particles. Although this discrepancy is quite large, the Langner et al. model does not include the detailed aerosol physics or chemistry necessary to properly delineate this problem . . . . Certainly, future gains in understanding the impact of anthropogenic sulfur on climate will require a better knowledge of aerosol particle formation.

[I]t is important that the relationships among anthropogenic sulfate, CCN (over a wide supersaturation spectrum), and N_d be understood for different cloud types. Because the influence of anthropogenic sulfate on the NH appears to be extensive, it is particularly important that these relationships be studied at a number of locations well distant from pollution sources.

Kaufman, Y.J., and D. Tanré (1994). Effect of variations in supersaturations on the formation of cloud condensation nuclei. *Nature,* 369, 45-48:

Sulfate aerosols can act as nuclei for cloud formation, thereby cooling the climate . . . ; however the magnitude of this effect is very uncertain . . . . Recently, Langner et al. calculated that at most 6% of the anthropogenic sulfur emission forms new particles, while 44% adds mass to existing sulfate particles activated in clouds. It was therefore suggested that previous studies . . . had overestimated the effect of sulfate aerosols on climate. Although it has been proposed that sub-CCN-size particles can grow to CCN-size in clouds . . . , this was thought to require the large supersaturations present in cumuliform clouds, rather than the smaller values characteristic of marine stratiform clouds, which are most important for radiative forcing. Here we show that natural variability of even low average supersaturations allows particles as small as 0.015 µm to grow to become CCN. This process can quadruple the CCN concentration and significantly increase the corresponding aerosol effect on climate.

There are many uncertainties in this result because of the complexities of heterogeneous sulfate chemistry and the possible competition of sulfates with other aerosol types, for example organic aerosol . . . . But it is our opinion that before the "details" of the model predictions (for example, variations in the supersaturation) are examined closely, it is too early to dismiss the importance of the effect of sulfate-cloud interaction on climate. It is possible that the models will not be able to assess this effect before enough measurements on the relationship between chemistry, aerosol size distribution, and CCNs are obtained.

Anderson, T.L., G.V. Wolfe, and S.G. Warren (1992). Biological sulfur, clouds, and climate. *Ency. Earth Syst. Sci. 1,* 363-376:

> Both the sulfur oxidation and new particle formation processes are poorly understood in the marine environment. Shipboard studies that have looked for increases in particle number, corresponding to increases in DMS levels, have been unable to observe such a correspondence consistently. (However, this measurement is difficult, and it is questionable whether it has yet been done properly.) In short, the current range of uncertainty spans the extreme possibilities that a given increase in the mass of nss sulfate will cause a proportional or a trivial increase in the number of CCN. Further tests of the relationship between nss sulfate mass and CCN number are therefore critical for assessing the sensitivity of the link between biogenic sulfur and climate.
>
> The proposal that marine biogenic sulfur influences global climate by controlling the number concentration of marine stratiform cloud droplets is thus enticingly plausible but far from established . . . . [T]he links connecting changes in DMS emissions to changes in cloud droplet number contain such large uncertainties that the possibility of very low climate sensitivity cannot currently be excluded. In addition, the effect that a change in "climate" . . . would have on the production of DMS by marine biota is completely unknown, even as to sign. Narrowing these uncertainties will require improved, global scale monitoring of parameters such as DMS concentration and CCN number, as well as intensive, process-oriented studies of the marine ecosystem and the cloudy marine atmosphere.

Quinn, P.K., D.S. Covert, T.S. Bates, V.N Kaapustin, D.C. Ramsy-Bell, and L.M. McInnes (1993). Dimethylsulfide/cloud condensation nuclei/climate system: Relevant size-resolved measurements of the chemical and physical properties of atmospheric aerosol particles. *J. Geophys. Res. 98,* 10411-10427:

> The ultimate goal of measurements such as those made here is to quantify the DMS/CCN/climate system under a variety of local physical and chemical conditions and to use the results as input into global chemical climate models. Therefore, the in situ measurement techniques used will have to perform with adequate sensitivity on the time scale of mixing within the marine boundary layer (~ an hour). These requirements apply to both gas and particle phase chemical measurements. As shown here, the simultaneous measurement of mass and number size distributions is useful in studying how the DMS oxidation products are distributed between particle production and growth. However, existing methods for measuring mass size distributions need to be modified or new techniques developed such that these size distributions can be measured on a shorter time scale and with a greater sensitivity.

Hoell, J.M., Jr., D.D. Davis, G.L. Gregory, R.J. McNeal, R.J. Bendura, J.W. Drewry, J.D. Barrick, V.W.J.H. Kirchhoff, A.G. Motta, R.L. Navarro, W.D. Dorko, and D.W. Owen (1993). Operational overview of the NASA GTE CITE-3 airborne instrument intercomparisons for sulfur-dioxide, hydrogen-sulfide, carbonyl sulfide, dimethyl sulfide, and carbon-disulfide. *J. Geophys. Res. 98,* 23291-23304:

> In summarizing these results, we note that the CITE 3 data base provides a comprehensive data base from which one may begin to analyze various sulfur budget issues. The data base confirms existing sulfur observations and theories while, at other times, provides emphasis to reconsider other issues. For example, the data clearly confirm (1) the importance of the ocean as a source of DMS and the rapid oxidization of DMS after transport from the marine mixing layer; (2) the continental source of H_2S, CS_2, and SO_2 as compared to a marine source; (3) the existence of a COS [carbonyl sulfide] latitudinal gradient (decreasing southward) estimated to be about 1.8 pptv/deg; (4) the higher concentration and variability of all sulfur gases in the NH, thus verifying the importance of NH anthropogenic emissions to global budgets; and (5) in the tropical Atlantic regions, devoid of major anthropogenic influences, photochemistry results in a net loss of ozone (3 to 5 pptv) during the day with a tendency for the marine mixed layer to be replenished at night via subsidence.

> On the other hand, the data base raises important questions. For example, (1) In terms of global sulfur budgets, is the transport and influence of NH air more important than originally viewed? (2) Is the COS budget complete, or are there missing source terms? (3) In view of the lower observed mixing ratios of H_2S and CS_2 in the tropical Atlantic (respectively, a factor of 3 and 10 lower than earlier data), what is the role and importance of the oceans as a source for these gases compared to continental sources? [Also: A host of improvements to instrumentation are needed.]

Lin, X., and W.L. Chameides (1993). CCN formation from DMS oxidation without SO_2 acting as an intermediate. *Geophys. Res. Let. 20,* 579-582:

> [T]o determine if such a chemical scheme does in fact exist [DMS → SO_3], more detailed studies on the kinetics of DMS oxidation under conditions typical of the MBL [marine boundary layer] are needed.

Kulmala, M., A. Laaksonen, P. Korhonen, T. Vesala, T. Ahonen, and J.C. Barrett (1993). The effect of atmospheric nitric acid vapor on CCN activation. *J. Geophys. Res. 98,* 22949-22958:

> Our simulations show that enhanced nitric acid concentrations can affect cloud droplet distributions by increasing the number concentrations and decreasing the mean size of the droplets. This can cause considerable changes to the radiative properties of low clouds. Furthermore, these effects can probably enhance cloudiness. First, it is likely that the smaller droplet size will decrease precipitation . . . so that the clouds will have

longer lifetimes. Second, the cloud formation can take place at smaller saturation ratios of water vapor. Third, with increased HNO_3 concentrations the disappearance of the cloud droplets due to evaporation is slower.

In marine regions the nitric acid concentrations are usually several orders of magnitude lower than those needed to significantly change the numbers of CCN, [but] other vapors [for example MSA (methanesulfonic acid)] may cause changes in CCN of marine clouds . . . . Further studies are needed to quantify the potential effect of the phenomena described in this study.

Lawrence, M.G. (1993). An empirical analysis of the strength of the phytoplankton-DMS-cloud-climate feedback cycle. *J. Geophys. Res. 98*, 20663-20673:

An understanding of the coupling between DMS flux and CN levels is tied up in the intricacies of the oxidation pathways leading from DMS to particulate sulfate. The examination of the long-term regional trends in DMS flux and CN measurements . . . revealed indications of a coupling, though it could not be discerned from the data whether the coupling was linear or nonlinear in nature. Since the nature of this relationship represented the most significant uncertainty in the model, . . . continued research into the mechanisms relating DMS emissions and CCN levels should help to reduce the uncertainties in the estimate. Such research should also help to illuminate the partitioning between homogeneous and heterogeneous oxidation pathways, which determines the relative roles of direct and indirect effects of enhanced aerosol and CCN levels.

Slinn, W.G.N. (1992). Structure of continental clouds before the industrial era: A mystery to be solved (Opinion). *Atmos. Env. 26A*, 2471-2473:

Which then raises a minor mystery: Is the creation of new CCN in the marine atmospheric boundary layer (MABL) negligible? If this investigation has been correct, most CCN are created in the free troposphere (FT), but that follows because the FT has ~10 times the volume of the atmospheric boundary layer (ABL). Yet, when the FT CCN mix back down to the ABL, there would be negligible simultaneous dilution of the FT's bank of CCN. So, where's the proof that most MABL CCN aren't created in the FT?

Clarke, A.D. (1993). Atmospheric nuclei in the Pacific midtroposphere: Their nature, concentration, and evolution. *J. Geophys. Res. 98*, 20633-20647:

Horizontal transects totaling over 35,000 km at about 9- to 10-km altitude [over the remote North and South Pacific] exhibited variability of approximately 3 orders of magnitude in both aerosol mass and number concentrations over spatial scales ranging from 1 to 1000 km. At these altitudes an approximate inverse relationship between ultrafine concentrations and the surface area of the larger aerosol was evident. Regions having lowest aerosol mass were characterized by aerosol thermal volatility indicative of a predominately sulfuric acid composition and by very high concentrations of ultrafine nuclei, indicative of recent homogeneous nucleation. These conditions were frequently observed but were conspicuously evident above clouds over the ITCZ [Inter-Tropical Convergence Zone]. The clean, free troposphere appears to be a significant source region for new tropospheric nuclei.

The relationship of atmospheric nuclei, CCN, and clouds is clearly a complex and coupled system. The importance of clouds to the global climate and the dependence of cloud properties on available CCN merits understanding of this system. The observations presented here demonstrate why boundary layer studies alone are unlikely to provide an adequate understanding of the system and make it clear that aircraft studies of the troposphere involving coordinated gas and aerosol measurements are essential. These studies should include deliberate efforts to assess the interdependent ways that gases, surface derived aerosols, existing aerosol surface area, environmental conditions, and clouds may interact to control both the CCN spectrum and related cloud properties.

Karl, T.R., G. Kukla, V.N. Razuvayev, M.J. Changery, R.G. Quayle, R.R. Heim, D.R. Easterling, and C.B. Fu (1991). Global warming: Evidence for asymmetric diurnal temperature change. *Geophys. Res. Lett. 18,* 2253-2256:

At present we lack an adequate understanding of the causes of differential changes in the mean and extremes of maximum and minimum temperatures. This is a fundamental characteristic of recent climate variation over a large portion of the NH land mass, and it must be better understood before we can confidently project the climate . . . . This will require rigorous atmospheric chemistry and climatological monitoring and analysis efforts. Additionally, improved modeling efforts are required which would consider the combined impact of changes of greenhouse gases, surface characteristics, and aerosols on the diurnal cycle in the atmospheric boundary layer.

Karl, T.R., P.D. Jones, R.W. Knight, G. Kukla, N. Plummer, V. Razuvayev, K.P. Gallo, J. Lindsey, R.J. Charlson, and T.C. Peterson (1993). A new perspective on recent global warming: Asymmetric trends of daily maximum and minimum temperature. *Bull. Am. Meteorol. Soc. 74,* 1007-1023:

Strong evidence exists for a widespread decrease in the DTR [daily temperature range] over the past several decades in many regions of the globe. There are many possible climatic factors that affect the DTR, but indications are that cloud cover, including low clouds, has increased in many areas that have a decrease in the DTR. The increases in cloud cover could be indirectly related to the observed global warming and increases of greenhouse gases, related to the indirect effects of increases in aerosols, simply a manifestation of natural climate variability, or a combination of all three.

A robust answer regarding the cause(s) of the decrease in the DTR will require efforts in several areas.

- First, an organized global effort is required to develop relevant and homogeneous time series of maximum and minimum temperature along with information on changes of climatic variables that influence the DTR (such as cloudiness, stability, humidity, thermal advection, and snow cover).
- Second, improvements in the boundary-layer physics and treatment of cloud within existing GCMs are critically important.
- Third, the treatment of both anthropogenic tropospheric aerosols and greenhouse gases must be realistically incorporated into GCMs with a diurnal cycle.
- Fourth, measurements need to be made to help clarify the role of aerosols.
- Finally, imaginative climate change detection studies that link the observed climate variations to model projections will be required to convincingly support any relation between anthropogenic-induced changes and the DTR.

Hansen, J., M. Sato, and R. Ruedy (1995). Long-term changes of the diurnal temperature cycle: Implications about mechanisms of global climate change. *Atmos. Res.*, in press:

Tropospheric aerosols alone cannot provide the continentally located forcing, i.e., they are not capable of damping the diurnal cycle as much as observed. Only an increase of continental cloud cover, possibly a consequence of anthropogenic aerosols, can damp the diurnal cycle by an amount comparable to observations.

Needed Observations.

The climatic implications of such cloud and aerosol changes can be discerned only if the changes are known globally on decadal time scales. The required long-term precision of measurements required to interpret decadal climate change are a cloud cover change [of] 0.4% as a function of cloud height ($\Delta p = 5$ mb) and a tropospheric aerosol optical depth change [of] 0.01 . . . . Although such precisions are not attained by existing or flight-scheduled satellite instruments, the capabilities have been demonstrated with relatively inexpensive long-lived planetary instruments. Specifically, a Michelson interferometer . . . has been shown to be capable of the cloud measurements in a multilayered atmosphere (Carlson, B.E., A.A. Lacis, and W.B. Rossow, 1993: "Tropospheric gas composition and cloud structure of the Jovian North Equatorial Belt." *J. Geophys. Res. 98*, 5251-5290). Similarly, a photopolarimeter has been demonstrated to be capable of such precise aerosol measurements (Travis, L.D., 1992: "Remote sensing of aerosols with the EOS Polarimeter," *Proc. Polarization and Remote Sensing, Vol. 1747*, 154-164, SPIE, San Diego), including accurate determination of aerosol and cloud microphysical properties. Without such data, long-term global climate forcings will remain unknown and interpretation and projection of any observed global climate change will be impossible.

Coakley, Jr., J.A., R.L Bernstein, and P.A. Durkee (1987). Effect of ship-stack effluents on cloud reflectivity. *Science* 237, 1020-1022:

> Because continental environments have significantly higher concentrations of CCN and lower relative humidities, a similar study should be undertaken to determine the effects of stack effluents from industrial centers on continental stratiform clouds.

Albrecht, B.A. (1989). Aerosols, cloud microphysics, and fractional cloudiness. *Science* 245, 1227-1230:

> Increases in aerosol concentrations over the oceans may increase the amount of low-level cloudiness through a reduction in drizzle—a process that regulates the liquid water content and the energetics of shallow marine clouds. The resulting increase in global albedo would be in addition to the increases due to enhancement in reflectivity associated with a decrease in droplet size and would contribute to a cooling of the earth's surface. [No recommendations are made, but there is the obvious complication for climate models: the need to predict CCN concentrations.]

Ackerman, A.S., O.B. Toon, and P.V. Hobbs (1993). Dissipation of marine stratiform clouds and collapse of the MBL due to the depletion of CCN by clouds. *Science* 262, 226-229:

> Marine stratiform clouds overlie about a third of the Earth's oceans and play a prominent role in the Earth's radiative heat balance .... It has been estimated that the global cooling that would result from a 4% increase in the area covered by marine stratocumulus would offset the expected warming from a doubling of atmospheric CO_2 concentrations ....
>
> Extensive sheets of marine stratocumulus are a climatological feature of the eastern regions of subtropical oceans, where the PBL [planetary boundary layer] is capped by a strong temperature inversion produced by large-scale subsidence. The vertical mixing that supplies moisture to these clouds and maintains the depth of the boundary layer is generally driven by cloud-top radiative cooling .... Clearly an understanding of the processes that determine the lifetime and albedo of marine stratocumulus is of critical importance to an understanding of the Earth's climate system.
>
> Albrecht . . . argued that increased CCN concentrations, which decrease cloud droplet sizes and reduce drizzle, can increase the fractional coverage of marine stratiform clouds because drizzle can regulate the liquid-water content during the lifetime of a cloud .... We have found, through numerical studies, that clouds themselves may reduce particle concentrations to such an extent that the clouds dissipate and, as a consequence, the boundary layer collapses.
>
> [No explicit suggestions, but obviously the theory should be tested, e.g., what entrainment of CCN from aloft?]

Baker, M.B., and R.J. Charlson (1990). Bistability of CCN concentrations and thermodynamics in the cloud-top boundary layer. *Nature 345,* 142-145:

> We . . . require simultaneous observational studies of the microphysical structure, albedo, and cloud-topped boundary layer structure, together with both background and anthropogenically enhanced CCN concentrations, to test the present model . . . .

Baker, M.B. (1993). Variability in concentrations of CCN in the marine cloud-topped boundary layer. *Tellus 45B,* 458-472:

> [The] eventual dependence of cloud cover on N [the CCN number concentration] is not exactly as envisaged by Albrecht (1989),[1] who suggested that the increased drying due to precipitation at low N might remove cloud. In fact, precipitation is self-limiting; when clouds get thin, the precipitation stops. [Our results] suggest that the link between cloud cover and N may result from the fact that N plays a role in controlling the heating profile in the cloud boundary layer. If this is correct, then those "equilibria . . . " that correspond to net in-cloud heating would in reality be short lived; ensembles of stratocumulus clouds might pass through these states, but only those clouds in which there is net cooling would be long lasting.
>
> A simple two-box model of the clear and cloud system suggests that the CCN number concentration in the clear part of the system is greater than in the cloudy part . . . . Measurements focusing on these points will be extremely useful in understanding the controls on cloud cover in the marine boundary layer.

Ackerman, A.S., O.B. Toon, and P.V. Hobbs (1994). Reassessing the dependence of CCN concentration on formation rate. *Nature 367,* 445-447:

> We find no evidence for bistability [in the CCN concentration]. However, we find that CCN concentrations are generally strongly dependent on their production rate, so that changes in the latter would influence the Earth's albedo.

Falkowski, P.G., Y. Kim, Z. Kolber, C. Wilson, C. Wirick, and R. Cess (1992). Natural versus anthropogenic factors affecting low-level cloud albedo over the North Atlantic. *Science 256,* 1311-1313:

> Cloud albedo depends on column-integrated liquid water content and the density of CCN . . . . A comparison of two independent satellite data sets suggests that, although anthropogenic sulfate emissions may enhance cloud albedo immediately adjacent to the east coast of the U.S., over the central North Atlantic Ocean the variability in albedo can be largely accounted for by natural marine and atmospheric processes that probably have remained relatively constant since the beginning of the industrial revolution.

---

[1]Albrecht, B.A., 1989. Aerosols, cloud microphysics and fractional cloudiness. *Science* 245, 1227-1230.

[No recommendations but some reader questions:

- These results are for low-level stratus (e.g., fog); what about for other-level clouds?
- What albedos for individual meteorological events?
- Correlation shown, but "correlation does not mean causation!"]

Kim, Y., and R.D. Cess (1993). Effect of anthropogenic sulfate aerosols on low-level cloud albedo over oceans. *J. Geophys. Res. 98,* 14883-14885:

[W]e examined satellite-measured low-level cloud albedo off the east coasts of North America and Asia at midlatitudes where anthropogenic sulfate sources are large and aerosols are transported eastward over the oceans by prevailing westerlies. The satellite data demonstrate enhanced cloud albedo near the coastal boundaries where sulfate concentrations are large. Similar trends are absent over ocean regions of the southern hemisphere that are removed from anthropogenic sulfate sources.

[No research recommendations, but their results stimulate some obvious questions from the reader:

- What influence from natural aerosol (from continents and from the "roaring forties")?
- What correlation with daily meteorology?]

Parungo, F., J.F. Boatman, H. Sievering, S.W. Wilkison, and B.B. Hicks (1994). Trends in global marine cloudiness and anthropogenic sulfur. *J. Climate 7,* 434-440:

A statistical analysis . . . shows a significant positive trend (4.2% increase from the 1930 baseline) in total oceanic cloud amount in the period between 1930 and 1981. The increase of total cloud amount for the NH (5.8%) was twice that for the SH (2.9%). The more consistent 30-yr (1952-1981) data show that the change in cloud amount (1952 base) was 1.5% for the globe, 2.3% for the NH, and 1.2% for the SH. The analysis shows that the greatest cloud amount increase was for altocumulus and altostratus clouds and that this increase was most pronounced at midlatitudes (30-50°N) . . . .

The trend and the pattern of cloud amount variations appear to be in accord with the temporal trend and geographic distribution of SO_2 emissions. It is hypothesized that sulfate particles, converted from SO_2, may modify cloud droplet spectra, causing affected clouds to be more colloidally stable than unaffected clouds. The longer residence times of affected clouds could cause increases of cloud frequency and cloud amount . . . .

Sinha, A., and K.P. Shine (1994). A one-dimensional study of possible cirrus cloud feedbacks. *J. Climate 7,* 158-173:

Early parameterizations [for climate models of the optical properties of cirrus clouds], in terms of ice water content, IWC (or path, IWP) of the radiative properties of cirrus clouds relied upon assumptions about the shapes and orientations of cirrus ice crystals. The differences between these assumptions and the great variety of actual sizes, shapes, and orientations of ice crystals inferred from aircraft observations was claimed to account for the persistent discrepancies between the observed dependence of the emissivity . . . and the shortwave albedo... on the IWC and the parameterized dependence . . . .

Aside from such theoretical obstacles, Stephens (1987) highlighted some of the problems regarding observational determinations of ice water content when he in turn ascribed the above discrepancies to the inability of measurements to account for the porosity of ice crystals. It has subsequently been found that perhaps the major hindrance to observational studies is the difficulty in obtaining reliable estimates of the numbers of small crystals (i.e., less than about 20 µm). These may play an important role in determining the optical properties of cirrus clouds . . . . Indeed small crystals may dominate the radiative characteristics of very high, cold cirrus . . . .

[T]he magnitude of the ice water content feedback is seen to depend substantially on the inclusion of the effects of ice crystals whose characteristic dimension is less than 20 µm . . . . This result underlines the need to develop instrumentation to accurately measure the number distributions of ice crystals in cirrus clouds within this size range.

Charlson, R.J., S.E. Schwartz, J.M. Hales, R.D. Cess, J.A. Coakley, Jr., J.E. Hansen, and D.J. Hofmann (1992). Climate forcing by anthropogenic aerosols. *Science* 255, 423-430:

Rates of new particle formation and of the time evolution of the size distribution of aerosols, which affect their optical and cloud-nucleating properties, are highly dependent on the rate of production of condensable material and on the concentration and size distribution of aerosol particles already present. A description of these rates is uncertain to orders of magnitude and is a major unsolved problem.

The indirect (cloud) forcing is more difficult to address than the direct forcing. Additional key phenomena that must be considered are (i) the relations between mass and number concentrations and composition of precloud aerosol concentrations and the number, concentration, and size distribution of cloud droplets formed in air containing the aerosol; and (ii) the dependence of aerosol perturbation to cloud radiative forcing on the nature of the cloud fields. The nature of the relation between CCN concentrations and sulfate concentrations, which is likely to be nonlinear . . . , is not established . . . .

The knowledge required to evaluate the influence of anthropogenic aerosols on cloud lifetimes, atmospheric dynamics, and the hydrological cycle includes the dependence of precipitation formation processes on cloud microphysical properties, such as number concentration and size distributions of cloud droplets. These secondary phenomena are considerably more difficult to quantify than the radiative properties, and additional research is required to accurately describe them in climate models.

Because anthropogenic aerosols are short-lived in the atmosphere and their sources are geographically localized, their concentration is not uniformly distributed. It is therefore necessary to specify pertinent aerosol properties as functions of location and time . . . . Vertical distribution information is required . . . . Horizontal distribution information is essential . . . . Temporal distribution information is needed . . . .

Better quantification of aerosol radiative forcing and of its relation to anthropogenic emissions requires globally representative measurements of aerosol and cloud properties. Long-term monitoring of the aerosol optical depth and of effective particle size can be achieved by satellites equipped with photopolarimeters . . . . Simultaneous monitoring of global cloud properties . . . . However, evaluation and improvements of inferences drawn from satellite measurements require that the measurements are tied to concurrent ground-based measurements of optical properties and to in situ surface and aircraft measurements of chemical and microphysical properties . . . .

The large uncertainties in the magnitude and geographical distribution of aerosol forcing . . . indicate the need for substantial research to improve the description of aerosol forcing. Among the components of such a research program would be a new set of satellites dedicated to questions of radiative forcing and able to provide the needed global coverage and high frequency of sampling. Connection of this global data set to local and column integral optical, chemical, and microphysical properties requires a set of ground-based observing sites carefully located to delineate the key characteristics of air masses that are either influenced or unaffected by anthropogenic sulfate. Understanding the chemical and physical processes that produce the aerosol and control its relevant properties requires both laboratory and field studies. These research activities must be integrated. Ground-based, airborne, satellite, and laboratory data acquisition activities should be coordinated, and modeling is needed at a variety of scales, from the mesoscale to the global scale, to aid in the design of measurement strategies, to provide a framework for data analysis, and to examine the climatic impact of anthropogenic aerosols.

Key questions remain regarding the climate influence of anthropogenic aerosol and its relation to the climate influence of increased concentrations of greenhouse gases: To what degree do such physically, altitudinally, geographically, and temporally different forcings compensate each other? Do such disparate forcings produce identical but opposite meteorological and climatological response? Have the radiative effects of sulfate aerosol decreased the magnitude of warming from anthropogenic greenhouse gases or delayed its onset or both? Or are there other climatic influences of anthropogenic aerosol that might have offset the radiative influences [of aerosols]? To what extent will reductions in SO_2 emissions for control of acid deposition exacerbate the greenhouse effect?

Penner, J.E., et al. (1994b). Quantifying and minimizing uncertainty of climate forcing by anthropogenic aerosols. *Bull. Am. Meteorol. Soc.* 75, 375-400:

> It is crucial that a research program be established to quantify the climate forcing by anthropogenic aerosols . . . . The prime components of such a program are (1) laboratory analyses and process studies, (2) surface based (land and ship) observations both for continuous and intensive periods, (3) airborne observations of key aerosol and chemical variables as well as radiation, (4) satellite observations of upwelling radiation and atmospheric extinction, and (5) models to connect estimates of aerosol sources to estimates of aerosol properties and to the geographically dependent observations.
>
> A research strategy has been developed here to quantify the direct forcing by anthropogenic aerosols. If aggressively implemented, this strategy promises to allow these effects to start to be incorporated into climate models within the next several years. To enable the development of a similar strategy for quantifying the indirect forcing by anthropogenic aerosols, a less well-structured program of basic research is required, wherein exploratory measurements of aerosol effects on clouds are made on a smaller scale. Much of the effort directed toward quantifying the direct effect of anthropogenic aerosols on climate forcing will benefit subsequent programs to quantify the indirect effects of anthropogenic aerosols.

Garratt, J.R. (1994). Incoming shortwave fluxes at the surface—A comparison of GCM results with observations. *J. Climate 7*, 72-80:

> The comparisons suggest that much of the excess surface net radiation found in GCMs [general circulation models] at continental surfaces (for 22 inland locations, by about 20 percent or 11 W m^{-2} on average, with significant seasonal and regional variations) is probably the result of excess incoming shortwave fluxes being calculated [by about 6 to 9 percent or 9 to 18 W m^{-2} on average (for 4 models: Commonwealth Scientific and Industrial Research Organization, Colorado State University, Geophysical Fluid Dynamics Laboratory, and U.K. Meteorological Office)] . . . . In the case of the incoming shortwave flux at the surface, observations over a period of time are affected . . . by [among other things] aerosol loading in the atmosphere . . . . Such effects are not simulated in most GCMs, so that there will be a tendency for model fluxes to be higher than those observed.

Karl, T.R., R.W. Knight, G. Kukla, and J. Gavin, 1995. Evidence for radiative effects of anthropogenic sulfate aerosols in the observed climate record. In *Aerosol Forcing of Climate*, R. Charlson and J. Heintzenberg (eds.). Dahlem Konferenzen 1994, Wiley, New York:

> The detection of an anthropogenic sulfate aerosol forcing on the surface air temperature is supported by a new analysis of daily surface air temperatures, cloud amount, and SO$_x$ emissions in North America. Within the contiguous U.S.A., SO$_x$ emissions increased by about 8 Tg between 1950

and 1970 and decreased by a similar amount from 1970 to 1990. The daily surface air temperature maxima under both clear sky and overcast conditions cooled by about 0.9°C from 1950 to 1970 and warmed by a similar amount thereafter. The results are consistent with the expected direct, and also possibly the indirect impact of SO_x emissions on surface air temperature .... The possible interference of factors affecting aerosol formation and transport, interaction with clouds, as well as the role of natural variations of atmospheric/ocean circulation require further assessment.

IPCC (1995). The effect of tropospheric aerosols. Excerpted from *Radiative Forcing of Climate Change*. The 1994 Report of the Scientific Assessment Working Group of IPCC, Summary for Policymakers, Intergovernmental Panel on Climate Change:

> New estimates of both the direct and indirect effect of anthropogenic aerosols in the troposphere have become available since IPCC [Intergovernmental Panel on Climate Change] 1992. In order to compare different aerosol effects it is useful to express them in terms of globally averaged values of radiative forcing. The direct radiative forcing due to increases in sulphate aerosol since 1850, averaged globally, is estimated to lie in the range -0.25 to -0.9 W m^{-2}. The direct effect of aerosol from biomass burning is estimated to lie in the range -0.5 to -0.6 W m^{-2}. Calculations of the indirect effect of aerosols are at an early stage. Preliminary results suggest that the radiative effect of aerosols on cloud radiative properties is probably a negative forcing and may be of similar magnitude to the direct effect. Note that in the global average the total aerosol induced radiative forcing is negative, but the absorption of solar radiation by carbonaceous aerosols may cause local positive radiative forcing. It is interesting to compare these estimates with the direct radiative forcing due to increases in greenhouse gases since pre-industrial times (+2.1 to +2.8 W m^{-2}), although it is unlikely to be appropriate to add the negative global radiative forcing of aerosols to the positive global radiative forcing of greenhouse gases.

B

Acronyms and Other Initials

ABL	Atmospheric boundary layer
ACE	Aerosol Characterization Experiment
ACE-1	Southern Hemisphere Marine Aerosol Characterization Experiment
ACE-2	North Atlantic Regional Aerosol Characterization Experiment
ACTM	Atmospheric chemical transport model
ADEOS	Advanced Earth Observation Satellite
AGCM	Atmospheric general circulation model
ARM	Atmospheric Radiation Measurement (Program)
ASTEX/MAGE	Atlantic Stratocumulus Transition Experiment/Marine Aerosol and Gas Exchange
AVHRR	Advanced Very High Resolution Radiometer
BASC	Board on Atmospheric Science and Climate of the National Research Council
CCN	Cloud condensation nuclei
CDNC	Cloud drop number concentration
CENR	Committee on Environment and Natural Resources
CFC	Chlorofluorocarbon
CLAES	Cryogenic Limb Array Etalon Spectrometer
CN	Condensation nuclei
COS	Carbonyl sulfide

DAEDALUS	Documenting Aerosol Electromagnetics, Defining Aerosol Lifetimes, and Understanding Sources (Research Stations)
DMS	Dimethyl sulfide
DOD	U.S. Department of Defense
DOE	U.S. Department of Energy
DOI	U.S. Department of the Interior
DTR	Diurnal temperature range
ENSO	El Niño/Southern Oscillation
EOS	Earth Observing System
EOSP	Earth Observing Scanning Polarimeter
EPA	U.S. Environmental Protection Agency
ERBE	Earth Radiation Budget Experiment
FT	Free troposphere
GAW	Global Atmospheric Watch
GCM	General circulation model
GHG	Greenhouse gas
GOME	Global Ozone Monitoring Experiment
HALOE	Halogen Occultation Experiment
HCFC	Hydrogenated chlorofluorocarbon
ICARUS	Interagency Climate-Aerosol Radiative Uncertainties and Sensitivities (Program)
IFN	Ice-freezing nuclei
IGAC	International Global Atmospheric Chemistry (Project)
IGY	International Geophysical Year
ILAS	Improved Limb Atmospheric Spectrometer
IPCC	Intergovernmental Panel on Climate Change
IR	Infrared
ISS	Integrated sounding systems
ITCZ	Intertropical convergence zone
IWC	Ice water content
IWP	Ice water path
LITE	Lidar In-Space Technology Experiment
MABL	Marine atmospheric boundary layer
MAC	Multiphase Atmospheric Chemistry (Activity)
MAGE	Marine Aerosol and Gas Exchange (Activity)
MBL	Marine boundary layer
MISR	Multiangle Imaging Spectroradiometer
MODIS	Moderate-Resolution Imaging Spectroradiometer
MSA	Methanesulfonic acid
NAPAP	National Acid Precipitation Assessment Program
NARE	North Atlantic Regional Experiment
NASA	National Aeronautics and Space Administration

APPENDIX B

NCAR	National Center for Atmospheric Research
NH	Northern Hemisphere
NOAA	National Oceanic and Atmospheric Administration
NSF	National Science Foundation
nss	Non-sea salt
NSTC	National Science and Technology Council
ONR	Office of Naval Research
PBL	Planetary boundary layer
PEM WEST	Pacific Exploratory Mission—West
PI	Principal investigator
PIXE	Proton-induced x-ray emission analysis
PM-2.5	Particulate matter of diameter less than 2.5 μm
PM-10	Particulate matter of diameter less than 10 μm
POLDER	Polarization and Directionality of the Earth's Reflectances
PSC	Polar stratospheric cloud
RFP	Request for Proposals
RH	Relative humidity
SAGE	Stratospheric Aerosol and Gas Experiment
SAM II	Stratospheric Aerosol Measurement
SCIAMACHY	Scanning Imaging Absorption Spectrometer for Atmospheric Cartography
SeaWiFS	Sea-Viewing Wide Field of View Sensor (ocean color satellite instrument)
SH	Southern Hemisphere
SPARCLE	Spaceborne Aerosols and Cloud Lidar Earthprobe
SURE	Sulfate Regional Experiment
TARFOX	Tropospheric Aerosol Radiative Forcing Observation Experiment
USDA	U.S. Department of Agriculture
USGCRP	U.S. Global Change Research Program
USGCRPO	U.S. Global Change Research Program Office
VOC	Volatile organic compound
WMO	World Meteorological Organization